Key Constellations

CALIFORNIA STUDIES IN MUSIC, SOUND, AND MEDIA

James Buhler and Jean Ma, Series Editors

1. *Static in the System: Noise and the Soundscape of American Cinema Culture*, by Meredith C. Ward
2. *Hearing Luxe Pop: Glorification, Glamour, and the Middlebrow in American Popular Music*, by John Howland
3. *Thinking with an Accent: Toward a New Object, Method, and Practice*, edited by Pooja Rangan, Akshya Saxena, Ragini Tharoor Srinivasan, and Pavitra Sundar
4. *Key Constellations: Interpreting Tonality in Film*, by Táhirih Motazedian

Key Constellations

Interpreting Tonality in Film

Táhirih Motazedian

UNIVERSITY OF CALIFORNIA PRESS

University of California Press
Oakland, California

© 2023 by Táhirih Motazedian

Library of Congress Cataloging-in-Publication Data

Names: Motazedian, Táhirih, author.
Title: Key constellations : interpreting tonality in film / Táhirih
  Motazedian.
Description: Oakland, California : University of California Press, [2023] |
  Includes bibliographical references and index.
Identifiers: LCCN 2022060393 (print) | LCCN 2022060394 (ebook) |
  ISBN 9780520382152 (cloth) | ISBN 9780520382169 (paperback) |
  ISBN 9780520382183 (ebook)
Subjects: LCSH: Motion picture music—Analysis, appreciation. | Tonality.
Classification: LCC ML2075 M66 2023 (print) | LCC ML2075 (ebook) |
  DDC 781.5/42—dc23
LC record available at https://lccn.loc.gov/2022060393
LC ebook record available at https://lccn.loc.gov/2022060394

32 31 30 29 28 27 26 25 24 23
10 9 8 7 6 5 4 3 2 1

CONTENTS

List of Figures and Tables ... vi
Acknowledgments ... ix
Preface—The Scope of This Book ... xi

1 The Theoretical Groundwork for Film Tonality ... 1
2 Tonal Analysis of the Soundtrack ... 27
3 Filmic Characters Rising Up and Settling Down ... 56
4 A Tale of Two (Tonally Symmetrical) Films ... 82
5 Unheard Sound Effects ... 108
6 Happy Accidents: Intentionality and Other Closing Thoughts ... 145

Appendix—Working Method for Creating a "Tonal Score" ... 157
Glossary ... 163
Bibliography ... 169
Index ... 179

LIST OF FIGURES AND TABLES

FIGURES

1 Sample tonal graph for *The Darjeeling Limited* 14
2 Sample tonal graph and tonal staff for *The English Patient* 16
3 Sample tonal staff levels for *The English Patient* 17
4 Sam's Khaki Scouts card (*Moonrise Kingdom*) 36
5 Tonal staff for *The Darjeeling Limited* 40
6 Mozart's parody of Salieri (*Amadeus*) 44
7 Salieri offers his prayer and God fulfills it (*Amadeus*) 46
8 Salieri's "Amen cadence" (*Amadeus*) 47
9 Salieri's unfulfillable longing—first example (*Amadeus*) 48
10 Salieri's unfulfillable longing—second example (*Amadeus*) 50
11 Salieri's yawns (*Amadeus*) 51
12 Salieri's murders (*Amadeus*) 52
13 Candle snuffed out by Salieri (*Amadeus*) 53
14 Sequence trajectories in *Persuasion* 60
15 E♭-major sequence in *Persuasion* 61
16 E-major sequence in *Persuasion* 61
17 Main theme in *Moonlight* 65
18 "BBB" theme in *Fantastic Mr. Fox* 77
19 Tom's realization in *The Talented Mr. Ripley* 87
20 Tom watches two coupled pairs in *The Talented Mr. Ripley* 90
21 Tom sings and dances in Dickie's clothes in *The Talented Mr. Ripley* 91
22 Transformation of E♭ keys in *The Talented Mr. Ripley* 92

23 Tonal graph for *The Talented Mr. Ripley* 96
24 Tonal staff for *The Talented Mr. Ripley* 97
25 Tonal graph for *The Grand Budapest Hotel* 103
26 Golden Ratio point in *The Grand Budapest Hotel* 105
27 Tonal symmetry in *The Grand Budapest Hotel* and *The Talented Mr. Ripley* 106
28 Two funerals in *The Darjeeling Limited* 116
29 Balloon theme in *Benny and Joon* 122
30 Squeaking cable car pulleys in *The Grand Budapest Hotel* 124
31 Conceptual parallel interrupted period in *The Grand Budapest Hotel* 125
32 Pitched birdcalls in *The English Patient* 127
33 Annotated score of "Bellbottoms" in *Baby Driver* 129
34 Annotated score of "Baby Let Me Take You" in *Baby Driver* 135
35 Tonal relationships in *Baby Driver* 141

TABLES

1 Britten works in *Moonrise Kingdom* 34
2 C♯-major and D-major cues in *The Darjeeling Limited* 39
3 Compositions in Mozart's portfolio, reviewed by Salieri (*Amadeus*) 49
4 Preexisting works in *Persuasion* 63
5 F-minor/major cues in *The Royal Tenenbaums* 70
6 A-minor/major cues in *The Royal Tenenbaums* 71
7 C-major cues in *The Royal Tenenbaums* 71
8 Final three cues in *The Royal Tenenbaums* 73
9 Theme transpositions in *Emma* 74
10 Important key areas of "Main" theme in *Emma* 75
11 "BBB"-based themes and their key areas in *Fantastic Mr. Fox* 80
12 Total musical time of all keys in *The Talented Mr. Ripley* 85
13 All D-minor cues in *The Talented Mr. Ripley* 86
14 Time periods, screen aspect ratios, and keys in *The Grand Budapest Hotel* 100
15 A-major cues in *The Grand Budapest Hotel* 100
16 C-major cues in *The Grand Budapest Hotel* 101
17 Sound effects during F-major sequence in *The Grand Budapest Hotel* 112
18 Sound effects during D-minor/major final showdown in *Fantastic Mr. Fox* 112
19 Tinnitus pitches in *Baby Driver* 133
20 Transposed keys of "Easy" in *Baby Driver* 140
21 Transposed keys of "Harlem Shuffle" and "Debora" in *Baby Driver* 140
22 Google Chat notification sound pitches 152

ACKNOWLEDGMENTS

I would like to offer my warmest and most heartfelt gratitude to the many people who made this book possible, of whom I have listed *a mere few* here:

To Iraj and Nourieh Motazedian, for leaving your homeland and forging a new life across the planet, so that I could have a chance in life.

To Robert Baldwin, for teaching me what to read and how to write, and for pushing me to develop into the person I wanted to be.

To Aida Baker, Rachelle McCabe, Kathryn Lucktenberg, Marlan Carlson, Judy Krueger, and Charles Creighton, for feeding me all the music my voracious appetite demanded. To Thomas Cockrell, for furnishing me with music so compelling that I quit my job at NASA to become a music theorist.

To Maureen Carr, for being my first champion in the field of music theory. To Kevin Korsyn, for his professional mentorship and advocacy. To Don Kinser-Traut, for introducing me to my first music theory research project. To Boyd Pomeroy, for helping me build a solid foundation of music theory skills, and for so many joyous and German-expletive-laced conversations about the music we love.

To Royal Brown and David Neumeyer, for kindly answering my questions and discussing my ideas when I first tiptoed into the field of film music theory. To Ronald Rodman, for providing me with the precedent and the encouragement I needed when I embarked on this research journey. To J. D. Connor, for being so generous with his time, counsel, and brilliant insights. To Rick Cohn, for teaching me all my favorite things in music theory, for motivating me with his fervent pursuit of new ideas, and for guiding me so wisely as I developed my own. To Dan Harrison, for *always* believing in me, for supplying the nurturing environment I needed to forge my own path, for setting a gracious example of the kind of professor I aspired to be, and for supporting me like a father when I lost mine during graduate school.

To Scott Murphy and Frank Lehman, for years of feedback, advice, and intellectual stimulation as we joyfully danced down the yellow brick road of film music theory.

To the Vassar College Dean of Faculty, Bill Hoynes, for providing the professional and financial support that allowed me to complete this book.

To Raina Polivka, for being the most supportive editor imaginable. To Jim Buhler, for his inspiring scholarship, enthusiastic guidance, and impeccable judgment as my series editor—Jim, I wrote this book with *you* in mind as my audience.

To Erica Stein, Erin Johnson-Williams, and Janet Bourne, for being my closest confidantes, compatriots, and cheerleaders in the world of academia.

And finally, to my beloved husband and favorite friend, Karl Hansen, who has accompanied me to the Moon and Mars and back again, who won my heart the moment our eyes met and whose blue eyes still make my heart skip a beat.

PREFACE

The Scope of This Book

Imagine a two-hour-long work in which keys hold rich associative meanings, relate to one another in hermeneutically significant harmonic relationships, and are temporally deployed in a symmetrical formation that reflects a programmatic theme within the narrative. This isn't a Wagner opera I'm describing—it's a contemporary, mainstream Hollywood film. Music analysts commonly use key-based approaches in studying symphonies and operas—so why not film? Why isn't key one of the standard parameters of film music analysis, as it is in most other genres of music?

Long-range tonality was declared early on to be irrelevant and unfeasible in the context of film music, and despite a few individual film analyses in the 1990s, the topic has remained dormant in its status quo. This book is the first systematic investigation into the arena of film tonality, and the results are fascinating beyond what one might have anticipated—revealing tonal constructs even Wagner would be proud to call his own. Using contemporary popular films as the context, I show how key and pitch analysis of a soundtrack can illustrate new layers of meaning and hidden insights within the filmic narrative. My approach considers the soundtrack in its entirety, including original scoring, preexisting music, pitched sound effects, and even instances of pitched dialogue. In addition to overturning the long-standing notion that key is an irrelevant parameter in the study of film music, this novel approach gives readers the chance to engage with some of their favorite films on an entirely new level.

I have written this book with the goal of being approachable for anyone with a basic understanding of music. To increase the accessibility for a wide range of readers, I have provided a glossary of both music and film terminology. So if you come across a term you're unfamiliar with, it is very likely defined in the glossary.

Chapter 1 lays out the premise, motivation, and logistics for this type of approach to the film soundtrack. Because soundtrack tonality is a new way of thinking about film, there are a number of questions people typically ask: Why and how should we consider a film soundtrack as a musical "composition?" What about all the nonmusical "silence" in a soundtrack? Does it matter that the music is composed and edited by many different people? Must the viewer be able to *hear* key relationships in order for them to matter? Do filmmakers plan key relationships across a soundtrack intentionally? The opening chapter answers these types of questions, priming and motivating the reader to understand and appreciate the film analyses in following chapters. This chapter also describes the different types of filmic tonal design elements, from their origins in Western art music.

Chapter 2 presents examples of film analysis, illustrating how tonal elements (presented in the previous chapter) can function in a soundtrack and contribute to the narrative.

Chapter 3 focuses on one particular tonal design element, to show how a parameter of film tonality can allow us to draw parallels between vastly different films.

Chapter 4 provides full-length film analyses of two seemingly unrelated films that are connected by an important element in the tonal layout of their soundtracks.

Chapter 5 explores the role of sound effects in film tonality, investigating how sound effects are pitched to interact with the music and the narrative.

Chapter 6 explores the question of intentionality in film tonality, briefly looks at tonal design in other media, and presents an overview of concluding thoughts about this new analytical approach.

The repertoire in this book is not drawn from any particular genre or time period (other than being, broadly speaking, mostly mainstream Hollywood films): they are films I happened to watch (for my own personal viewing) in which some tonal feature struck my attention, spurring me to investigate further. I specifically did not make any attempt to limit my corpus to "the best" films or composers, because I want to demonstrate that meaningful tonal features can be found in *any* type of film setting. The hope is that after reading this book, you will start listening to key and pitch relationships in everything you watch—and even in the world around you too. You will begin noticing that your sonic environment is teeming with pitched sound effects and music that interact in fascinating and beautiful ways, when you allow your ear to hear them on the same aural plane.

As for the book's title, *Key Constellations*, I elucidate this term in chapter 1 after having laid out the theoretical groundwork for my analytical approach—so I invite the reader to read the first chapter with that anticipation in mind.

A supplementary website houses many of the film excerpts discussed in this book, as well as additional score excerpts and film stills; throughout the text the

(✶) symbol indicates the presence of supplementary materials on the website. I urge readers to watch the film excerpts even if you have seen the films before, because these annotated clips allow you to internalize the tonal techniques discussed in the book. Access this supplementary website at:

motazedian.net

1

The Theoretical Groundwork for Film Tonality

We are about to embark on a new and unconventional approach to film music. Before we launch into the details of this approach, let's step down the aisle and take our seats for a couple of films already in progress, where we will trace the paths of two protagonists who have very little in common with one another:

The first is Anne Elliot, the downtrodden heroine of *Persuasion*, a 1995 film adaptation of the eponymous Jane Austen novel. Societal conventions and familial pressures have relentlessly silenced Anne until she has forgotten that she even has a voice. Having been coerced into rejecting the man she loves, she settles into a subdued shell of her former self, and the years teach her to silently accept her unhappiness. When fate brings Captain Wentworth back into her life (miraculously still single), Anne is too enervated to overcome the inertia of her resignation and respond to his renewed interest. But his proximity reawakens her confidence, and by the end she finally makes the bold (and culturally shocking) move to defy her friends and family and seize her own happiness.

The second protagonist is the eponymous hero of *Fantastic Mr. Fox*, a 2009 stop-animation film featuring a cast of animals and three evil farmers. Mr. Fox has far too much confidence and audacious energy. His reckless actions cause him to endanger the safety of his community and lose the respect of his family, so he must wage war against the farmers to save his friends and redeem himself. This requires Mr. Fox to slow down, think carefully, and act prudently—basically suppress all his instincts and start behaving like an adult. He struggles against this at first, but eventually he matures and manages to unite the animals in a victorious battle against the farmers.

So why are we putting a zany animal caper and a British period drama side by side? The motivation for this unlikely pairing lies in the auditory realm: in each case, the tonal layout of the soundtrack (the overarching arrangement of keys) reflects the dramatic circumstances of the protagonist. The soundtrack for *Persuasion* features a large number of classical piano works, as is common in Jane Austen adaptations. But what is unusual here is that not a single one of these works (by Bach and Chopin) is presented in its original key: every one of them has been transposed to new keys. Starting at the beginning of the film, each musical composition is systematically lowered a half step below its original key. Then, at a certain point in the film, the music is transposed a whole step *above* its original key. Through the first three-quarters of the film the pieces lowered in pitch correspond with Anne's depressive state. When Anne finally rises up to reclaim control of her life, the key of the music likewise rises upward. Thus the soundtrack depicts Anne's character arc *tonally* by reflecting the stages of her life journey (in even finer detail, as we discover in chapter 2). The overall tonal trajectory of the film, too, begins in A major and ends in B major, delineating Anne's narrative trajectory by means of "directional tonality."

Such directional tonality is also at play in *Fantastic Mr. Fox*, but in the opposite direction. Unlike sluggish Anne, rambunctious Mr. Fox must grow more subdued in order to become the best version of himself (he must settle down in order to grow up). Thus the soundtrack in this film features a *downward* shift—beginning in E major and ending in D major. The trajectory for the "happy endings" in these two films move in opposite directions, and so, too, do the tonal trajectories of their soundtracks. On the one hand, Mr. Fox begins his journey from a state of manic hyperactivity and must calm down in order to attain happiness. Anne, on the other hand, begins from utter calm and lethargy, and she must grow more animated and proactive in order to achieve her happy ending. Thus, the tonal direction for Mr. Fox is to settle down and for Anne to rise up (a whole step). We might disregard such tonal configurations as mere coincidence were it not for the conspicuous transposition of preexisting music, which is almost certainly deliberate. Transposition on the local level and directional tonality on the global level turn out to be rather common, and we will see the same devices in such films as *Emma* (1996) and *The Graduate* (1967) when we explore this technique in greater depth in chapter 2.

These analytical snapshots suggest that key can be an important consideration in film, so that raises the question of why key is ignored as a significant parameter in film music analysis. After all, key is one of the basic building blocks of music and a central property of a work's musical identity, and analytical attention is routinely given to key in most other genres of music. (We say "Beethoven's Fifth Symphony in C Minor," not "Beethoven's Fifth Symphony in Four Movements and Thirty-four Minutes" or "Beethoven's Fifth Symphony with the Very Iconic Motive.")[1] So *why not film?* The answer to that question is rooted in certain theoretical notions formed in the context of classical music, which initially posed some ideological hurdles in

---

1. Thank you to Scott Murphy for this droll idea.

the very different context of the film soundtrack—for an overview of this history, see my earlier work (Motazedian 2016: 2–25). But by now, well into the twenty-first century, we are certainly ready to adapt and broaden our ideas about how large-scale tonality can function in new settings. So let's explore the five main theoretical issues that will pave the path for our pursuit of film tonality.

1. What Does Film Tonality Entail?

I have coined the term *film tonality* to refer to the large-scale arrangement of keys of all musical entities in a film soundtrack—including original scoring, preexisting music, and pitched sound effects and dialogue. There are existing terms for large-scale tonal organization, of course, but because film soundtracks have many unique considerations, it's useful to have a term specifically for this context. But let me begin by tying in and clarifying some of the relevant terminology. The term *tonal design* has long been used to refer to large-scale key organization in musical compositions. To differentiate between two possible modes of organization, David Beach (1993) draws a distinction between the terms *tonal structure* and *tonal design*. *Tonal structure* captures the hierarchical relationship of pitches within a single key (in a Schenkerian sense), while *tonal design* refers to a deployment of keys not necessarily governed by a global tonic and possibly influenced by extramusical factors. *Tonal structure* is best suited for monotonal, monopartite (single-movement) contexts; it is not optimized for dealing with directional tonality, double-tonic complexes, associative tonality, and other tonal practices of the late nineteenth century and beyond, nor does it account for tonal development across gaps such as breaks between movements. These factors make tonal structure a less appropriate model for explaining large-scale key relations across multipartite works. *Tonal design*, however, is analytically descriptive rather than prescriptive (in the sense of not presupposing a global tonic or functional harmony) and is therefore capable of depicting any manner of tonal deployment. The versatility of this approach makes tonal design a better tool for analyzing expansive, multifarious works like opera and film, which do not adhere to standard musical forms and do not necessarily conform to monotonality.[2]

A few scholars in earlier decades considered the idea of large-scale tonality in film from a tonal structure approach, and understandably this ill-fitting Procrustean bed didn't produce compelling results in the context of film soundtracks.[3] What we take away from these earlier inquiries is the importance of acknowledging that tonal organization in a film will behave differently than tonal organization in a Mozart piano sonata—just as Mozartian tonality behaves differently than

---

2. See Motazedian (2016: 4–10) for a synopsis of the debate over tonal design in opera.

3. See Motazedian (2016: 10–14) for an overview of Adorno and Eisler (1947 [2005]), Cochran (1990), Neumeyer (1998), and Neumeyer and Buhler (2001).

Mahlerian tonality. Tonal design in a film soundtrack is not bound by the type of harmonic logic (especially the assumption of functional monotonality) we might encounter in classical repertoire. Indeed, even *within* classical repertoire there is no single harmonic logic that neatly codifies hundreds of years of Western music—so expecting a film soundtrack to exhibit tonal behavior akin to a sonata's seems wholly unreasonable. Fruitful analysis of film tonality thus requires a flexible perspective and openness to broader definitions of tonality.

Whereas tonal structure entails a *prescriptive* approach in which we look for what *should* be happening (i.e., how a single tonality organizes the music into a structure) and conform the work to the methodology, tonal design entails a *descriptive* approach in which we look at what *is* happening and adapt the methodology to the work. Through this approach, tonal idiosyncrasies provide a rich resource for dramatic interpretation, freeing the viewer-listener to follow their analytical instincts, inspired by the narrative context.[4] With this approach in mind, let's consider an abstract example of how tonal structure and tonal design might function differently in the context of analyzing a film soundtrack: a film beginning in C major and ending in F♯ major would be deemed highly anomalous from a tonal structural standpoint, impelling us to interpret this harmonic anomaly as a dramatic anomaly. However, there could be a narratively cogent reason why the film begins in C major and ends in F♯ major—for example, these keys might be associatively paired with the characters who appear in the opening and closing scenes, respectively.

In such a context the design approach would allow C major to move to F♯ major without raising an analytical eyebrow, whereas the structure approach would spur the analyst to conjure an aberration in the narrative (because this harmonic motion would be seen as I–♯IV, which is aberrant in a monotonal system). Imposing the structural value judgment of I–V–I onto a system that is not operating under the requirement of I–V–I is like using a German grammar book to grade an English paper: different system, different rules. There may be shared elements and origins and traceable influence, but the two systems nevertheless function in fundamentally distinct ways.

2. *Must We* Hear *It for It to Be Valid?*

Long-range tonality—in music in general, not restricted to film music—has often been questioned on the basis of audibility: Must a listener *aurally* perceive tonal relationships for them to "matter?" Musicians have long debated this question without reaching a consensus. Allow me to adapt terminology from Carolyn

---

4. For excellent related discussions of "dramatic tonality" in opera, see Latham (2008), Bribitzer-Stull (2006b), McCreless (1982), and Katz (1945). For an overview of tonal design in opera, see Motazedian (2016: 4–10).

Abbate's 2004 discussion of *gnostic* and *drastic* forms of perception to characterize the two sides of the argument, where *drastic* perception is sensory and immediate, and *gnostic* perception is intellectual and mediated. On the *drastic* end of the spectrum, some believe that key relations only matter if the keys are directly contiguous and perceived immediately and naturally (i.e., without *trying* to hear them). Others do not require keys to be immediately adjacent but still feel that key symbolism (such as associative tonality) is dependent upon aural perception—which essentially restricts it to those possessing absolute pitch.

Along those lines, some scholars contend that keys cannot carry associative meaning across works or repertoires (since the vast majority of listeners do not have the capacity to *hear* them). On the *gnostic* end of the spectrum, Nicholas Cook (1987) assigns the task of tonal perception to the mind rather than the ear. Based on his experiment (in which listeners report their perceived sense of coherence in works whose endings are recomposed to different keys), Cook asserts that "the tonal unity of a sonata is of a conceptual rather than perceptual nature, in contrast to the directly perceptible unity of a single phrase" (1987: 204).[5] The conceptual nature of long-range tonality described by Cook can be discerned gnostically, even if it cannot be perceived drastically. Thus gnostic perception is a more useful tool for exploring tonal relations in the context of large-scale works.

A gnostic approach helps us address one of the main issues of tonal audibility: the discontinuous nature of the film soundtrack. The long expanses of non-musical sound occupied by dialogue and other sounds (let's call them "gaps") that can separate music cues certainly do prevent us from being able to *hear* the connection of one key to another. But the lack of drastic perception does not invalidate the need for gnostic investigation. Requiring listener perception as a prerequisite is problematic in *any* repertoire: by this logic, only those possessing absolute pitch would find the key of C minor relevant to an analytical understanding of Beethoven's Fifth Symphony.[6] As for the ability to hear tonal relations across gaps, are we meant to *hear* (on a drastic level) the systemic prolongation of D♭ over the course of the four-day gap between the closing D♭ of *Das Rheingold* and the closing D♭ of *Götterdämmerung* in the *Ring* cycle?[7] And if we cannot hear it, does that nullify Wagner's conscientious tonal design? Composers such as Wagner and Shostakovich use key associations over the gap of different works (and across the

---

5. Robert Gjerdingen (1999: 164–166) poses a fair critique of Cook's scientific methodology in this experiment but says that he does still find Cook's conclusion to be "persuasive" (164).

6. Also, consider that even within the traditional analytical realm of classical music, many forms of analysis present information that is not drastically perceived by the listener. For instance, most listeners cannot *hear* the completion of rows in a twelve-tone composition, but they can understand them gnostically, with the aid of analysis.

7. Bribitzer-Stull aptly questions "the perceptual limits of Schenkerian theory" by asking, "can we really hear prolongation over any span of time?" (2006b: 330).

gap of years), and while listeners cannot drastically hear these connections, we can gnostically understand them.[8]

Thus gaps do not present a barrier to long-range tonal analysis.[9] In the context of film soundtracks, Scott Murphy even proposes that "with this approach to tonal unity, a musical score chopped up into cues (and other types of self-containers) that are further separated by considerable stretches of time becomes a strength instead of a weakness."[10] This is because each musical cue is usually short enough to remain monotonal (as compared to longer, continuous musical works in which modulations and tonicizations can make it difficult to define which key we are "in" at any given moment). And let us not forget that the visual structure of film is *also* inherently fragmentary, but film viewer-listeners have long been accustomed to "connect[ing] the dots" across disjunctions in filmic form (Rodman 2010: 168). We can apply this same logic to the *sonic* structure of film. Filmic narrative is likewise filled with gaps (we encounter flashbacks and flashforwards and ellipses without any sense of cognitive disjunction), and once we start analyzing, you will find that musical gaps can interact with the inherently disjointed structure of film in surprising meaningful ways.

3. How Can a Soundtrack Be a "Composition?"

It is natural to wonder how a collection of cues in disparate musical styles and written by different composers can cohere together to form a unified "work." Michel Chion provocatively asserted "there is no soundtrack," meaning that audio elements are more strongly bonded to the image than they are to one another (1994: 40). But this notion has been challenged on a number of points. The term *mise-en-bande* (coined by Rick Altman, McGraw Jones, and Sonia Tatroe [2000] as a parallel to *mise-en-scène*) posits the soundtrack as a unified entity, just as multipartite and multilayered and "constructed" as its visual counterpart. James Buhler points out that the soundtrack, like the image track, derives its "power . . . not from some mystical unity but from the way *editing* (in particular) productively structures the tensions among the various components" (2001: 55–56).[11] Thus, even

---

8. For further discussion in favor of the gnostic appreciation of key relations, see McCreless (1996: 106–108).

9. If we *were* to consider gaps a deal-breaker, this would negate tonal relationships between movements of a multipartite work like a string quartet (especially since it is now common practice for performers to retune their instruments between movements, thereby severing the aural continuity of one movement's ending key and the next movement's starting key). Using multipartite works as an analogy, cues in a film can be treated like the movements of a multimovement work. (And the logic of key relations across movements is rarely one of a straightforward tonal structure in Beach's sense of the term.)

10. Personal e-mail communication on 9 January 2015.

11. Buhler goes into even greater depth discussing Chion's infamous dictum in his 2019 book (249–256).

though the dialogue, music, and sound effects are edited and spliced together, we can acknowledge the resulting end-unity of the soundtrack, the same way we accept the end-unity of the image track.

In recent years film music scholars have been making a concerted effort to knit together the customarily segregated components of the Hollywood soundtrack: for instance, David Neumeyer (2015) presents a model for analyzing the soundtrack as a whole, and Danijela Kulezic-Wilson (2016, 2017, and 2020) depicts the soundtrack as a "composition of speech, music, and sound effects" (2020: 17).[12] Tonal analysis of the soundtrack is another natural step in this modern trend of recognizing the soundtrack as a coherent *work*.

4. Who Is the "Composer" of a Soundtrack?

If we treat the soundtrack as a composition, who is its "composer"? Because so many people are involved in the production of a soundtrack, it is not possible to attribute every aspect of it to any one person. Is it rational to consider the creation of multiple artists as one unified work? Film scholarship settles this question with the notion of auteurism, which ascribes authorship of a film to the director, as the person whose artistic vision shapes the "contributions [of the entire team] into a whole" (Bordwell and Thompson 1997: 38). In the music realm, lieder, operas, and ballets present an obvious answer to the question of composite authorship, being works that require the collaboration of multiple artists. While these artforms require the cooperation of different *types* of artists (e.g., a composer and a poet), there are also many precedents for the *same* types of artist (e.g., two painters) collaborating to produce a single work. Beatles songs,[13] Diabelli's *Vaterländischer Künstlerverein* variations,[14] and *pasticcio* operas[15] are well-known examples of collaborative musical efforts. Andy Warhol and Jean-Michel Basquiat famously collaborated on a series of paintings over several years, and Robert Rauschenberg and Jean Tinguely worked jointly to create sculpture. And the field of architecture

---

12. Neumeyer and Buhler similarly state that "if music is a structuring of sound in time, as many twentieth-century aestheticians have claimed, then conceptually the *mise-en-bande*, with its complex interplay of music, dialogue, ambient sound, effects, silences, and so forth, can be understood as a kind of musical 'composition'" (Buhler in Neumeyer 2015: 100).

13. Contrary to the popular view that John wrote the words and Paul wrote the music, both artists have (on numerous occasions) averred that they collaborated more or less equally in all aspects of the composition.

14. Anton Diabelli invited fifty-one composers to contribute variations on his simple waltz theme, resulting in a theme and variations published in 1823–24.

15. This use of the term *pastiche* refers to a medley-type work compiled from different sources, not an imitative work in the style of another artist, work, or period.

involves collaboration as a matter of course.[16] When multiple artists contribute to a composite work, even in the absence of a single autocratic "Creator" with an all-encompassing vision, the resulting artwork acquires its own gestalt.

Composite composition is the standard procedure in film music, making authorship exceedingly difficult to pinpoint. Even when a film score is attributed to a single composer, the reality is often far more complex than the traditional model in which a composer wrote the themes and arrangers orchestrated them. This model has increasingly been superseded by a more collaborative working method like the one epitomized by Hans Zimmer and his Remote Control Productions (RCP) studio, whose stable of composers collaborate closely on all projects. Zimmer is often criticized for creating a factory line of composers who churn out his musical bidding ("corporate classicism" as Nicholas Reyland [2015] calls it), but Nicholas Kmet (2018) gives a more nuanced account of the working environment at RCP. Kmet depicts an artist collective in which individual composers cooperate and interact with one another on their separate projects, and he paints Zimmer less as a boss than as a "producer and purveyor of film scoring talent" who enjoys surrounding himself with compatriots and collaborators (2018: 6).

Kmet (2018) describes a "degree of freedom" in which "responsibilities are less strictly apportioned" and "roles less delineated" than critics might expect. Furthermore, he points out that due to Zimmer's liberality in doling out music credits and his relative unconcern with claiming them for himself, it can be rather futile to name a specific individual responsible for what ends up in a film score. The collaborative working model is ever more common as a cost-efficient method of film scoring, better suited to handle the enormous scope and punishing deadlines of the industry. In addition, as Casanelles (2016) and Kulezic-Wilson (2016, 2017, and 2020) have recently documented, non-composer members of a film's music department have increasingly *creative* roles in the process, actively generating musical/sonic content—which complicates the question of authorship even further when we consider the entire soundtrack (rather than the original scoring in isolation). Thus attribution of a film composer's (or director's) name is merely a synecdoche of the entire group of artists responsible for creating a film soundtrack (or the whole film itself).

The bottom line is, regardless of the creative process or working method or precise allotment of contributions, we, the viewer-listeners, can perceive and interpret the resulting gestalt of a composite creation as a coherent and unified artwork.

---

16. In addition to the dozens or hundreds of architects working at a single firm, it is common for even the most prominent firms led by so-called "starchitects" to partner with other firms on large-scale projects.

5. Did They Do It on Purpose?

This is an irresistible inquiry when one sees how intricate and thoroughgoing a film's tonality can be. But the question of intentionality requires a bit of theoretical unpacking, to clarify our analytical priorities and the best way to achieve them. Analytical approaches can be broadly categorized as poietic and esthesic.[17] With *poietic* analysis the reader interprets a work according to what the author intended, and with *esthesic* analysis the reader interprets a work using whatever factors seem fruitful or relevant. Both approaches have their merits, of course, but the esthesic approach tends to be most productive in the context of film. As Ronald Rodman astutely observes, "film has become one of the rare American cultural artifacts in which esthesic readings have revealed a masterwork greater than the sum of its parts"—and thus a film "must be 'read' by the viewer for it to make sense" (1998: 128). Literary theorists have long espoused the notion of reader-response criticism, in which meaning is not created by the author or embedded within the text but generated by the reader upon interaction with the text.[18]

Regarding analysis as a *reading* rather than a *revelation* (of authorial intention) empowers the analyst to move beyond the limitations of what the author may have intended. Navigating authorial intent is a tricky business, and the problematic nature of authorial self-analysis is well-known and widely discussed.[19] And in the context of film, with its multiple layers of components woven together by multiple teams of people, it can be difficult (and possibly artificial) to attribute specific artist intentions at a level of granularity higher than the aforementioned synecdochic-type attribution.[20] But from an esthesic perspective, we can interpret the features in a film regardless of who created them or whether their creation was "deliberate or adventitious" (Keller 2006: 85).[21]

---

17. These terms originate in the context of semiotics: *poietic* indicates production (author-based), and *esthesic* indicates reception (reader-based). As forms of analysis, these two terms capture what the author intends and what the reader perceives, respectively. Consult Nattiez (1990) for further discussion.

18. For more on reader-response criticism, consult the works of Roland Barthes (1977), Wolfgang Iser (1978), Stanley Fish (1980), Michel Foucault (1979), and Alexander Nehamas (1986).

19. Wimsatt and Beardsley (1946) set the pond on fire with their discussion of "intentional fallacy," and literary theory has long embraced this concept. With specific regard to music, see Richard Taruskin (1996, 2000, 2004, and 2009, passim) and David Latham (1997 and 2008).

20. See Schatz (2010) for a convincing account of how filmic meaning is created at the industrial level.

21. We can further deconstruct the notion of authorship by invoking the distinction between *writer* and *author* formulated by Foucault (1979) and Nehamas (1986). The *writer* is the human being who created the work, while the *author* is "whoever can be understood to have produced a particular text as we interpret it." The *author* is "produced through an interaction between critic and text" and is "postulated to account for a text's features." Through this lens, the *writer* "exists[s] outside their texts" and therefore has "no interpretive authority over them"—and is even prone to "misunderstand[ing] their own work" (Nehamas 1986: 686). In the context of film, where there can be hundreds of *writers*

Taking a reader-based approach to analysis does *not* mean that we must shun filmmaker intentions—only that we should (1) take them with a grain of salt, and (2) treat them as *one* of many possible factors contributing to our interpretation of a film. As we will see in chapter 5, filmmaker commentaries can augment and complement our filmic interpretations in interesting ways. Of course, a reader-based approach is highly subjective—but really, *all* analysis is subjective at its core. The notion of "objective" analysis is somewhat of an illusion, as every analytical decision—from choosing which elements to analyze, to selecting which methodology to apply—is intrinsically an interpretive action.[22] Depending on one's analytical goals, this inherent subjectivity can be considered an asset rather than a liability.[23] Embracing analysis as a creative, interpretive art form, "the perception of a poetic work resides in the (active) making of another poetic work" (Lewin 1986: 385).[24] From this perspective, the analyst's task is not to uncover the artist's intentions or reveal the one "correct" interpretation but to create a reading—one of many possible readings—that will be compelling and engaging to other readers.

As for the term tonal *design*, this word does not necessarily imply the premeditated actions of a "designer." Let's think of tonal design as the tonal landscape of a work and follow the landscape analogy: whether a specific person (e.g., landscape architect) deliberately planned it this way, or whether it ended up this way adventitiously through a variety of factors (e.g., weather, geology, insects, animals, gardeners, etc.), the resultant end-product can be referred to as the "design" of the landscape. Indeed, even *with* the deliberate planning of one specific landscape architect, all those other factors would still influence the resulting design.

Allow me one more nature-based analogy in elucidating the title of this book, *Key Constellations*: celestial constellations obviously weren't "designed" for our

---

involved in creating a single work, the designation of *author* is especially pragmatic. Barthes states that "the birth of the reader must be at the cost of the death of the Author," because "a text's unity lies not in its origin, but in its destination" (1977: 148). If we apply Nehamas's "writer/author" nomenclature to Barthes's "death of the Author" proclamation, we might declare that the *writer* is dead while the *author* is alive and well. However, readers are accustomed to shifting seamlessly between differing ascriptions of agency without cognitive disjunction (Monahan 2013), so perhaps it is not necessary to kill the *writer* either. We can ascribe a creation to "Wes Anderson," without having to pinpoint *which* Wes Anderson—*writer* or *author*—is responsible for what.

22. Even within rule-governed methodologies such as serial analysis and Schenkerian analysis, the selection of "important" pitches can be regarded as an interpretive act. Consult Latham (1997 and 2008), Guck (2006), Lewin (1986), and Agawu (2009) for discussions on the fundamentally interpretive nature of analysis.

23. See Temperley (1999) for a thoughtful perspective on the different goals of musical analysis.

24. Or, in the classic words of Harold Bloom: "The meaning of a poem can only be another poem" (1973: 94).

astronomical purposes, they are our own interpretive constructs.[25] Nevertheless, they are an important part of our engagement with the night sky, despite the fact that they are entirely subjective (existing only from our unique physical vantage point) and selective (the constituent stars bearing no proximity to one another)—in other words, entirely reader-based. Key constellations in a film soundtrack have a much higher level of designed-ness than star constellations, obviously, but the point is that our engagement with constellations does not depend on this criterion. Pattern recognition has always been a crucial part of human functioning, and we constantly seek to understand the world around us by engaging with patterns. Thus let us derive enjoyment and value from the key constellations of a film just as we marvel at the glittering formations of our night sky.

A NEW APPROACH

In forging this new approach to film tonality, let me reiterate the core guiding principles. This approach is fundamentally descriptive (rather than prescriptive) in that we allow the particulars of each film and its narrative to determine the most fitting analytical methodology (rather than being dictated by the precepts of an established methodological system). Therefore, we think of tonality in terms of "design" rather than "structure," since we are not bound by the assumptions of monotonality. This does not mean that all elements of structural tonality must be excluded from soundtrack analysis—but the foremost motivation for defining hierarchical relationships between keys in a soundtrack must be the filmic (narrative) context.

To form a complete picture of a film's tonality, we consider the entire soundtrack—the *mise-en-bande*—rather than extract a single element from its sonic context and examine it in isolation. Considering the complete soundtrack, however, does not imply that we must include or explain each and every sonic element present in the film.[26] Analysis in general (musical and otherwise) requires the analyst to judiciously decide which elements to consider to generate a compelling, constructive interpretation. Therefore, we need not assume that every instance of music and sound in a film *must* be meaningful to the overall tonal design.

Our tonal exploration of a medium as complex as a film soundtrack is greatly enriched by going beyond simply what our *ears* perceive and building our interpretation through cumulative analysis. Each time we reanalyze a film, we accrete additional insights to our interpretation, resulting in a deeper, more

---

25. The term *key constellations* is borrowed from Ronald Rodman's (2011) filmic tonal design context, who borrowed it from Scott Balthazar's (1996) operatic tonal design context.
26. As Lehman astutely says, "an attitude that demands every harmonic event be explicable forces irregular key choices to become 'anomalies' requiring an 'answer' or solution" (2012b: 100).

nuanced understanding of the film. That deepened understanding shapes our subsequent viewing experiences regardless of whether or not its component insights can be *heard* by the unassisted ear. Insofar as our goal is an esthesic, reader-based interpretation, intent is not required as permission or "proof" of our reading. This is not to say that composer/director commentary should be excluded, but rather, regarded as simply one possible element in our interpretive toolbox.

The most enticing incentive for tonal analysis is the interpretive power it yields. This approach provides a new perspective for interpreting the narrative—and for comprehending the narrative work of music—as it allows weaving in fascinating layers of meaning that sometimes substantiate and sometimes subvert one's previous understanding of a film. And it also allows the viewer-listener to engage with film on a new level. It enables us to perceive the soundtrack as a cohesive entity rather than a patchwork of discrete components. We can use tonality as a tool for exploring the gestalt of a film's sonic plane (and its relation to narrative structure), much as we might assess the visual plane in terms of elements like cinematography or set design.

LOGISTICS

Now that we have addressed the conceptual concerns relating to film tonality, let's address the logistics involved in undertaking this type of analysis. To fully fathom the tonal landscape of a film, we must consider the tonal elements of the complete soundtrack: this means original music (composed for the film), preexisting music, pitched sound effects, and pitched dialogue—really anything on the soundtrack that can be construed as bearing tonal implications. A great deal of scholarship on film music focuses solely on original scoring, but this music does not exist in a sonic vacuum: a Beatles song playing diegetically on a jukebox, a Vivaldi concerto playing nondiegetically over a montage sequence, an ice cream truck jingle playing metadiegetically through a character's mind—any and all music heard during the course of a film is part of its tonal *mise-en-bande*. So, too, are pitched sound effects (bells, sirens, phones, clock chimes, etc.) and pitched dialogue events, such as a character whistling or humming or delivering their line in a singsong manner. Anything that strikes your ear with a distinct sense of pitch should be noted.

Since the soundtrack is the only true and complete record of a film's sonic content, one's ear is the main authority for this type of analysis. Written film scores (in addition to being difficult to obtain) account for only a fraction of the soundtrack because they do not include preexisting music or sound effects. They might not even accurately reflect the original music heard in the film, since any number of changes may have occurred post-production. Therefore, we must create a kind of alternative-format "score" (think of it as a tonal screenplay, if you will) for the entire film soundtrack, before we can begin our analysis. The appendix provides detailed descriptions of the working method for creating a "tonal score."

Once we have constructed our tonal score for the film soundtrack, analysis can begin. As we established earlier, our goal is to generate a musical analysis that interacts with the filmic narrative in a meaningful way, so how do we go about determining which keys are "meaningful" in a film? Certain characteristics become apparent during the construction of the tonal score (for instance, observing that more than half the cues are in F minor), and these provide a starting point for analytical inquiry. But for the most part, the real work begins after the tonal score is complete, as we search for salient schemes or patterns. For example, if the film's narrative is structured around three climactic events, and G major accompanies each of these climaxes, we would want to investigate this key further. If we wish to explore the associative power of G major, our first step would be to verify that the three iterations of this key are not simply repetitions of the same cue, which would collapse the distinction between theme and key and nullify its power of signification. For tonality to be demonstrated to be "an independently productive parameter," we must be able to separate tonality from thematicism (Lehman 2012b: 110). (For example, Simon & Garfunkel's E-minor "Scarborough Fair" is repeated five times in *The Graduate* [1967], but this *by itself* is not enough to designate E minor as an associatively significant key, since there is no way to distinguish between repetition of theme and repetition of key. We would need at least one other cue in E minor—or other factors—to initiate the associative value of the key as independent from the song.)[27]

The next step is to scrutinize if and how G major is used elsewhere in the film. If G major is *not* used elsewhere (and the three iterations consist of distinct musical works), we might probe the precise nature of this key's role in the narrative. If G major *is* used in a variety of other instances throughout the narrative, this key may not have a specific function in the film—but we must evaluate numerous other considerations before disregarding it. For instance, other characteristics might be common to all these instances of G major. Perhaps this key always occurs in the context of one particular character, location, time period, or plot motif. Perhaps cues in G major always precede cues in C minor, and the former functions as a dominant springboard for the latter. Perhaps every instance of G-major music in this film is metadiegetic and conveys the internal struggles of its characters. Or perhaps G major is the exclusive key of the film's nondiegetic music, and

---

27. There *are* other factors that collaborate to imbue significance to the key of E minor in *The Graduate*, as we will see in chapter 2.

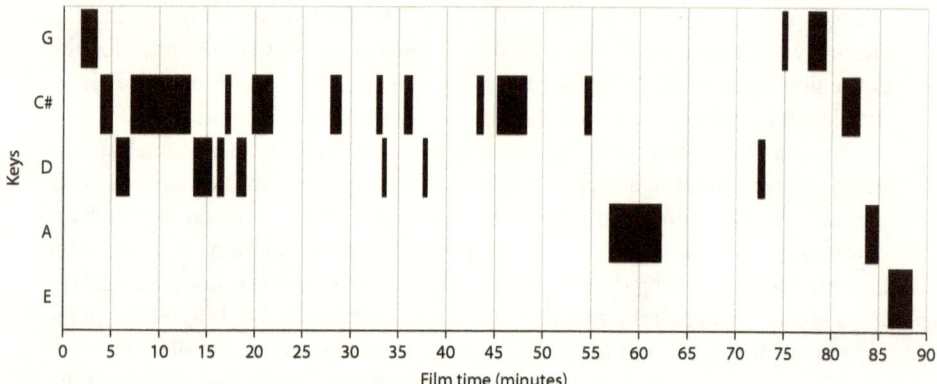

FIGURE 1. Sample "tonal graph" for *The Darjeeling Limited*.

the key functions as a sort of Greek-chorus commentary. There are any number of possibilities. The point is we have many perspectives from which to approach the assessment of a key's role in a film, involving different factors, and a thorough analysis will consider these factors (and the ways they interact with one another) from every angle. Sensitivity to the specific context of each film is imperative for getting the most out of this type of analysis.

Sometimes it can be helpful to visually explore the information from a tonal score by arranging it graphically. Rather than using existing musical graphing systems, which come with inherent ideologies and assumptions built in, I have devised a graphing method (adapted from a Gantt chart, a type of bar chart commonly used for scheduling) for mapping out tonal design. This "tonal graph" makes it easier for us to study the tonal layout of a long, complex work like a soundtrack. Figure 1 provides a sample graph for *The Darjeeling Limited* (2007), a film we explore in greater depth in chapter 2. The X axis represents time, and each musical cue is represented by a rectangular bar, the horizontal width of which is determined by the length of the cue in minutes. The Y axis separates keys[28] into their own rows, the vertical ordering of which can be customized to reflect whichever factor seems most relevant in the context: in order of pitch, chronological entrance, or relative importance of keys.[29] (In this particular case, keys in *The Dar-*

---

28. Modal distinction is collapsed according to the Schoenbergian concept of twelve chromatic keys (rather than twenty-four major/minor keys) to keep the graph as visually clear as possible.
29. The criteria for "importance" can be based on a key's role in the narrative, or the total cumulative amount of time spent in a key, or (in many cases) a combination of both. In many films, the keys with the most significant narratival roles are also the keys in which the most time is spent. As is the case

jeeling Limited are ordered along the Y axis according to chronological entrance.) Horizontal gaps in the graph, in which no other cue or key is represented, account for either musical silence (i.e., the absence of music) or time spent in "unimportant" keys (narratively and durationally inconsequential enough that they are not labeled or represented on the graph).

Tonal graphs can help reveal patterns and properties that might not otherwise be apparent. For instance, looking at the tonal graph for *The Darjeeling Limited* the most salient feature is the alternation of step-related keys: musical cues in this soundtrack mainly alternate between C♯ and D major. The film's narrative is centered around conflict and tension, and the tonal tension of the stepwise relationship reflects the conflicted nature of the relationships between the characters. The music fluctuates by step as the characters continually confront one another, take sides and switch sides, fight and make up, oscillate between trust and mistrust. Once the tonal graph makes us aware of this stepwise conflict, we start to notice that the entire soundtrack is peppered with a plethora of step-related sound effects, maintaining a haze of tonal tension in the sonic atmosphere. This goes on for the first two-thirds of the film, and once the key of A major is achieved (around minute 60), the incessant stepwise juxtaposition of both music and sound effects ceases. A major is the key during which the main characters finally achieve the resolution they have been so desperately seeking, and they can now move forward in peace. And as a token of their deliverance from strife, A major returns at the end of the film (as the characters ride off serenely into the sunset) to lead into the final cue in E major, creating a IV–I plagal motion that bathes the characters in the tonal absolution of the "Amen cadence."

Tonal graphs can also allow us to generate another useful type of diagram. Figure 2 (top) shows how keys are used associatively in *The English Patient* (1996), with three-fourths of the soundtrack's musical time spent in the keys of B minor and G major. B minor is the key of the protagonist's tragic, mysterious journey, and G major represents the love story. The keys of D major and B♭ major also carry their own associations (explored in chapter 2), but they are far less important in the tonal design and narrative of this film. We can use a musical staff to depict this information in a slightly different (and more familiar-looking) format in Figure 2 (bottom). This "tonal staff" conveys the relative importance of keys through the size and duration of their note heads (with horizontal placement corresponding to the X axis of the tonal graph).[30] Using an adaptation of an adapted method, we can represent this film's tonal design at "foreground," "middleground," and

---

with many other analytical methodologies, context is crucial in determining which factors are important and relevant to one's analysis.

30. This is *not* a Schenkerian graph, despite the vague resemblance.

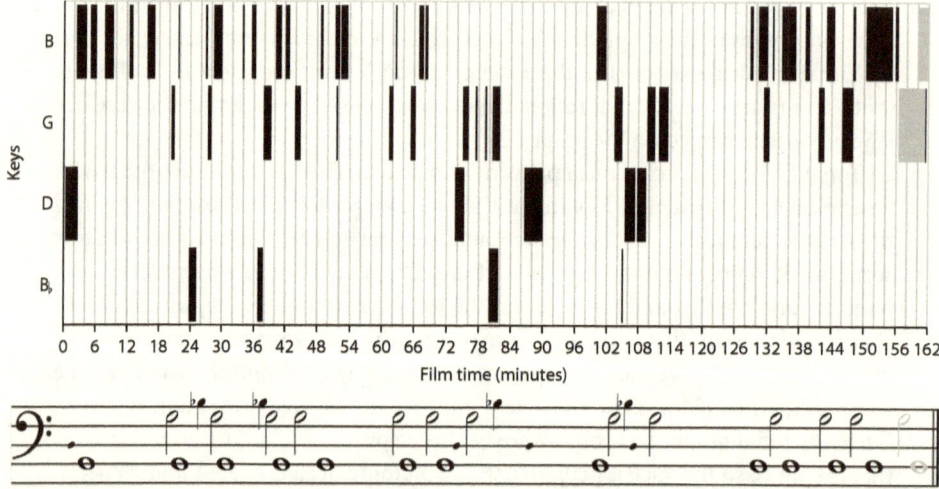

FIGURE 2. Two different types of tonal diagrams for *The English Patient*.
Top: "Tonal graph" for *The English Patient*. The gray bars and noteheads indicate music that occurs during the (closing) credits.
Bottom: "Tonal staff" for *The English Patient* corresponding to the tonal graph above (with gray noteheads corresponding to the gray bars).

"background" levels (Figure 3) to illustrate the most fundamental key relationships.[31] To generate the middleground level, we strip away the less important key of D major and some of the less significant back-and-forth oscillations between B minor and G major. To generate the background level, we exclude B♭ major and boil down to the most narratively significant B minor–G major oscillations (the context of which is explained fully in chapter 2). The importance of B minor in *The English Patient* is apparent in Figure 2 and Figure 3: this key on its own accounts for almost half of the soundtrack's total musical time, the film begins and ends in B minor, and most of the main events occur in this key. G major seems like a secondary key, in comparison (which is why the tonal staff designates B minor with whole notes and G major with half notes). But if we delve deeper into the connection between narrative and key (which we do in chapter 2), it becomes possible to interpret the key of G major as the core kernel of the soundtrack. The soundtrack features many G-major cues (both preexisting and original), but we can trace the presence of G major in this film back to one source: the Aria of Bach's *Goldberg*

31. I have adapted McCreless's (1990) non-Schenkerian adaptation of Schenkerian levels, in which he (McCreless) proposes a compromise between Schenker's strict monotonality and Leo Treitler's associative tonality.

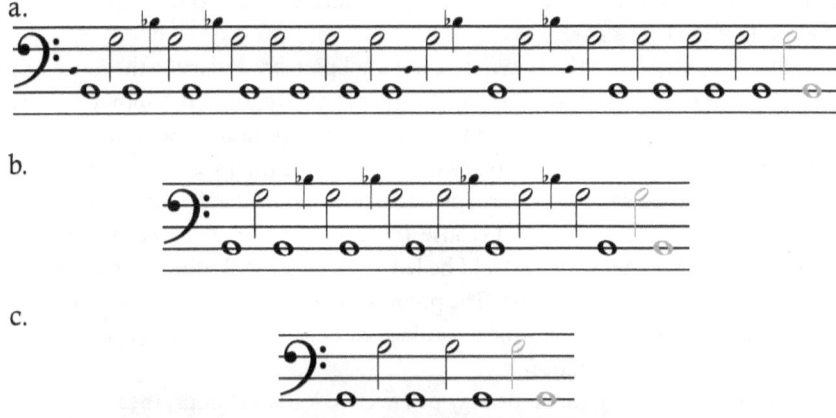

FIGURE 3. Tonal staff represented at foreground, middleground, and background levels for *The English Patient*.
a. "Foreground" level
b. "Middleground" level
c. "Background" level

Variations. Director Anthony Minghella selected this piece as the starting point for the soundtrack, even before filming began.[32] The Bach is used diegetically to mark a significant event in the narrative, and composer Gabriel Yared used it as the compositional model for his G-major love theme—so the key of G major in this film germinates from the Bach Aria.

Building from the vantage point that G major came first (in this soundtrack), let's examine how B minor relates to it. Musically speaking, B minor arises from a *Leittonwechsel* transformation of G major, and the *Leittonwechsel* has been consistently used in modern American film to convey "the contemplation of some kind of considerable loss, commonly the death of a loved one" (Murphy [2014c: 487] and [2014b]). Since G major is the love-story key and B minor is the key of the tortured remembrance of that love, B minor can be said to derive from G major (via *Leittonwechsel* transformation) in this film, reflecting the "loss of a loved one" film-music trope. If we extend this tonal realization (that B minor is *caused by*

---

32. Minghella remarked: "When I'm writing I'm already marinating the film in some kind of musical landscape, and I bring those elements to Gabriel [Yared] as a gift. So for instance, in *The English Patient*, I came to him with a sense of how Bach was a very strong ingredient" (Bernard and Khanna 2007).

G major) to the narrative, we realize that protagonist's entire predicament occurs as a result of his love affair.[33]

From this perspective the romance is not a subplot, secondary to the more conspicuous plot elements of war and espionage and exploration, but rather, the primary story. Perhaps, then, the B-minor and G-major note-head values on the tonal staff (see Figure 3) should be switched, because the love story initiates the mysterious-journey story? Or perhaps both keys should have whole-note status (as in a double-tonic complex) because the two story lines are so inextricably intertwined? Or perhaps they should be left as is, because B minor is still the key of the ostensible "main" story line? The point is that versatility and flexibility in our analytical representations allow us to explore, hone, and communicate the subtleties of our filmic interpretation.

While these graphing methods may be effective for some films, their use is neither rigid nor requisite (we will explore a number of analytical techniques in the remainder of this chapter). Because film tonality is so variable, our methods need to be adaptable to the unique context of each film. In some contexts, for example, the graph may be more productive if keys are arranged (along the Y axis) according to their chronological entrances rather than relative importance.[34] In other contexts, graphs may not contribute any additional insights to the analysis and are thus superfluous. Judicious selectivity and assessment are therefore inherent features of this kind of work—as they are with analysis in general. We cannot weight every component equally but must be critical in deciding where to place emphasis in order to produce a compelling interpretation. It's important to resist the urge to include every musical element or concoct convoluted rationales for components that defy characterization. Not every key will be significant to a film's tonality, and as discussed above, a successful analysis need not strive for complete tonal accountability.

While we can perform this kind of tonal analysis on any film, certain types of films lend themselves more readily to it. Contemporary, mainstream, narrative films tend to yield the clearest results because of the prevalence of compilation scoring in these films. The variegated nature of the contemporary compilation soundtrack (featuring a pastiche of different composers, works, and genres) makes it easier to avoid conflating tonal and thematic association (as discussed earlier). And the discrete start- and stop-times of musical cues in these types of films provides a clearer setting for parsing tonality. In a classical Hollywood score (such as Max Steiner's *Gone With the Wind* or John Williams's *Star Wars*), music is virtually continuous throughout the film, which makes it difficult (and at times, impossible) to delineate discrete cues or key areas. Within long stretches of through-

---

33. We might even see a reflection of this causality in the closing credits, which begin in G major and end in B minor.
34. We'll see a great example of this in chapter 3, in the analysis of *The Grand Budapest Hotel*.

composed underscoring, the music may modulate casually (i.e., without clear boundaries or specific harmonic purpose), making it problematic to define when we are actually "in" a new key.

Films with wall-to-wall scoring also spend a great deal of time in ambiguous harmonic territories, spinning out amorphous music beneath the dialogue (similar to the nonspecific music that accompanies operatic recitative), and these kinds of passages obscure our ability to designate one particular key. Another model of film scoring that does not fit comfortably within the scope of tonal analysis is the popular compositional style based on triadic chromaticism (as epitomized by James Horner, Jerry Goldsmith, and John Williams, among others), which features a constant shifting of the tonal center, making it difficult to establish a single "tonic" for a given passage or cue. This style of scoring is more transformation-oriented than tonic-oriented, so it becomes artificial to designate the key of a particular cue. Thus these types of films are better-suited to methodology that emphasizes localized transformations rather than long-range tonal design.[35]

SURVEY OF TONAL ELEMENTS

Tonal design is an umbrella term that can encompass a variety of different components in different film settings. Many of these techniques are also found in nineteenth-century music and opera, but I briefly explain how they work in the context of film soundtracks. I have grouped them into three categories:

1. Elements that rely on assigned meaning.
2. Elements that rely on harmonic relationships.
3. Elements that use keys abstractly.

(Note that I use the terms *technique*, *element*, and *component* somewhat interchangeably, to avoid the undue implication of intentionality sometimes associated with the term *technique*.)

1. Elements that Rely on Assigned Meaning

In this category of techniques, the meanings of keys are established within the context of a film. In other words, there is no inherent meaning in the use of a particular key outside the specific context of that film.

Associative tonality is the most ubiquitous form of filmic tonal design. This term was coined by Robert Bailey in the context of opera, to describe a referential system in which a particular key is consistently paired ("associated") with a

---

35. Consult the writings of Scott Murphy (2006, 2014b, 2014c, and 2023 forthcoming) and Frank Lehman (2012a, 2012b, 2013b, and 2018) for fascinating, in-depth studies of transformational analysis of film music.

particular extramusical or dramatic element. For example, the key of C major might be used for the protagonist, or D minor might be used to convey a depressive state, or E major might be used during excursions to a particular location. Robert Bailey (1977), Siegmund Levarie (1978), Matthew Bribitzer-Stull (2001, 2006b, and 2012), and others have written about the use of associative tonality in the operatic works of Wagner and Verdi.[36] Given the formal, logistical, and aesthetic similarities between opera and film, it is unsurprising to find this technique being used in film as well. In film associative keys are most often used to represent characters and psychological/emotional states, and sometimes structural non-narrative elements of the film. All constituents of the soundtrack (original music, preexisting music, pitched sound effects, and pitched dialogue) can be used to contribute to the establishment of associative networks.

In further support of associative tonality, the **transposition** of music (and sound effects, which we explore in greater depth in chapter 4) to different keys can help reinforce established tonal schemes. Leitmotifs in Wagner's *Ring* cycle, for instance, are transposed to different pitch levels to coordinate with the overarching associative network. Transposition can also create subtle shades of meaning by bringing the associative references of different keys into dialogue with one another, either through the superimposition and comingling of the two associations or through the supplanting of one by the other (Bribitzer-Stull calls this "associative transposition," in the context of Wagnerian opera).[37] Transposition in film soundtracks can be especially intriguing if it involves preexisting music, since we might wonder why the effort was made to transpose an existing commercial recording into a new key—either through reperformance or through pitch manipulation of an existing recording.

I have coined the term **tonal wink** to describe the playful correspondence between the letter name of a key/pitch and the first letter of a character, place, or object name. For example, the key of D major used as the associative key for a character named Dominic (or the city of Durango) might be deemed a tonal wink. Examples of these types of musical cryptograms abound in classical music, such as J. S. Bach's use of the B-A-C-H motif (B♭-A-C-B, translated from its original German nomenclature) and Dmitri Shostakovich's use of the D-S-C-H motif (which translates to D-E♭-C-B). And closer to the film industry, the famous NBC chime features pitches G-E-C, which tonally wink at the acronym for its original corporate parent, the General Electric Company.

Another witty element, but one that operates at a deeper level of narrative involvement, is **tonal agency,** in which filmic characters can exert their agency by producing particular keys that bear the (associative) power to change their for-

---

36. See also Marcozzi (1992), Hepokoski (1989), Balthazar (1996), and Latham (2008).
37. Bribitzer-Stull (2001: 161–172 and 2015: 195–200) and Gauldin (2015).

tunes. This phenomenon tends to occur in the context of characters who are manipulative and controlling, and who use music as one of many methods of control. Of course, this line of thinking brings up questions of how agency might be designated or delineated between filmmakers and film characters, and the nuances of agency can be explored from a variety of perspectives in such sources as Levinson (1996), Hartley (2009), Winters (2010), and Monahan (2013).

Expanding beyond "the world of the work" (Monelle 2000: 12), some films feature a tonal design that is built around the key of another work, for which I have coined the term **intertextual tonality**. In these films a particular preexisting work is preselected for a particular purpose in the narrative, and a network of cues are built around the key of that work. This type of tonal reference creates intertextual meaning in the film by importing the array of meaning and context surrounding that preexisting work.[38] That preexisting work's key may not have had any associative connotations in its original context, but the new context (the film) utilizes its key as a synecdoche for that preexisting work's content or character. From there, other keys may be selected on the basis of evocative harmonic relationships to the original key, and thus an associative network (or "constellations of keys," to quote Balthazar [1996: 78] and Rodman [2000]) is built up around the key of a preexisting work—forming the tonal starting point for the soundtrack. There are countless precedents for tonal intertextuality in classical music, as, for example, demonstrated by Patrick McCreless (2010) in the context of Shostakovich and the key of D minor. Intertextual analysis entails exploring "the hidden roads that go from poem to poem" (Bloom 1973: 96; cited in Korsyn 1991: 9).

Tonal pairings entail the coexistence of two more or less equally weighted keys that alternate throughout the course of the work. Tonal pairings and their more controversial cousins, double-tonic complexes (in which two third-related keys fuse together to form a four-note sonority that governs as *one tonic*), have been widely discussed and debated in the context of nineteenth-century music,[39] and the discussion continues today in the arena of rock music.[40] The general consensus is that although one key may be more prominent than the other at a given moment (Bailey 1985: 122), they are equal partners in creating a balanced narrative based on "the tension between two tonal centers" (Kinderman 1980: 106). Tonal pairing works smoothly and uncontroversially in film soundtracks, since there are no conventional requirements for structural monotonality. And tonal pairing is an especially fruitful harmonic scheme in a film context, because of the rich potential for dramatic capacity inherent in the coupling or conflict of two keys.

---

38. For excellent discussions of intertextuality in music, see Korsyn (1991) and Klein (2005).

39. See Bailey (1969 and 1985), Kinderman (1980), Krebs (1981 and 1996), Bribitzer-Stull (2006b), BaileyShea (2007), for example.

40. See Nobile (2020), Richards (2017), Schultz (2012), and Spicer (2017), for example.

A film whose soundtrack is heavily centered around two prominent keys (such as a tonal pairing) may revisit those keys during the closing credits to create a kind of **tonal coda.** In music a coda designates "parageneric space" (Hepokoski and Darcy 2006: 282) that conveys an "after-the-end" quality (Caplin 1998: 179) through thematic means—that is, thematic material that lets you know you're in the end zone. In the context of film tonality, where we don't take thematicism into account, a tonal coda designates a framing function in the soundtrack that is external to narrative space proper. The closing credits provide this parageneric space wherein the main keys of narrative space can be revisited, not only with themes and cues heard during the film but often with new material.

The use of **singular keys** (i.e., a key used only once in a film) is a soundtrack feature to be approached cautiously, because of the earlier proviso that an associative key be represented by multiple cues. But some films do present a compelling case for contemplating the meaningful (associative) use of singular keys. A singular key can highlight the importance and uniqueness of a particular narrative event, for example, and might be paired with other distinctive features (such as the sonic foregrounding of a volume spike and/or some sort of contrasting visual effect) to further draw the viewer-listener's attention. The case for a singular key is especially compelling if it is the only one in a film—that is, all other keys used in the soundtrack occur in multiple different cues, with just the one single key reserved for a particular moment. Thus the multiple-cues stipulation does not automatically or unconditionally rule out singular keys—but there should be a cogent and convincing narrative motivation in order to consider overriding it.

2. Elements that Rely on Harmonic Relationships

The techniques in this category tap into existing meaning, established outside the context of a specific film, associated with commonplace harmonic relationships. The significance we attach to these ubiquitous key relationships (e.g., parallel major/minor) in Western music has developed through hundreds of years of convention.[41] The **parallel relationship** is among the most basic key relations, preserving the tonic and changing the mode. This relationship allows for an easy juxtaposition between happy/sad, good/bad polarities, especially obvious because of the inherent sense of contrasting character we read into a single tonic being deflated to minor or boosted to major.

The **relative relationship** likewise changes the mode but also changes the tonic (in order to preserve the key signature). Like the parallel relationship, relative-related keys can portray dramatic dichotomous pairings, though not as aurally direct of a comparison, due to the different tonic. The preserved key signature

---

[41] See Bribitzer-Stull (2015: 170–175) for an excellent discussion of the parallel relationship in nineteenth-century musico-dramatic settings.

provides the potential for a slightly more nuanced reading: if key signature can be taken as the metaphorical DNA of a key, we might, depending on the film's narrative context, interpret relative-related keys as two halves of a Janus-faced entity.

The **Leittonwechsel relationship** is another diatonic third-relation, like the relative relationship but with root motion in the *opposite* direction. Speaking in terms of a chord progression, the *Leittonwechsel* replaces a chord's tonic with its leading tone (e.g., the C in a C-major triad is replaced with B, so that a C-major chord becomes an E-minor chord). A staple of nineteenth-century triadic harmony, the *Leittonwechsel* has become one of the favorite harmonic devices in Hollywood film music during the past few decades. Murphy (2014b) brilliantly characterizes the prevalent use of the *Leittonwechsel* transformation to convey the contemplation of loss of a loved one in film music, in the 1980s and onward. This progression has been used so habitually in film and television music that it eventually began to signify sadness in general and has now progressed to serve as a powerful and ubiquitous signifier of emotion *in general*. At this point, even casual viewer-listeners are well conditioned to the emotional affect implied by this harmonic signifier—to such a degree that the *Leittonwechsel* can now be used ironically or humorously, as evidenced in the 2020 McDonald's french fry commercial, "Trust Me."[42] (Also, I guarantee you a laugh if you ask people to imagine Homer Simpson staring longingly at a jelly donut, while playing a *Leittonwechsel* oscillation at the piano.) And although this harmonic relationship is normally deployed at the localized chord-to-chord level, we can extrapolate its signification to the broader key-to-key level: *Leittonwechsel*-related keys in a film soundtrack can be used to portray loss and intense emotion, especially if each of the keys has associative value.

Like the *Leittonwechsel*, the **stepwise relationship** draws its association from the localized level—but in this case the pitch-to-pitch interval level (rather than chord-to-chord). Because major and minor seconds are considered highly dissonant intervals in Western music, our musical convention has long equated harmonic seconds with tension and strife. Step-related keys, likewise, can be used to portray conflict in film, depending on how the keys are associated—for example, by pitting one key/character against another.[43] Alternately, the stepwise relationship can be used to convey a sense of movement in the narrative, based on the upward or downward motion between two keys. Bailey (1977) describes stepwise key relations in Wagner operas (using the term *expressive tonality*), equating upward motion with increasing dramatic tension and downward motion with

---

42. View the McDonald's "Trust Me" commercial (produced by the Wieden+Kennedy ad agency) with its humorous *Leittonwechsel* motif here: https://vimeo.com/445022398.

43. Rodman (2000) explores the character relationships represented by step-related keys in the 1937 film *Maytime*.

decreasing dramatic tension.[44] And pop music and related genres (stretching back to the jazz era) have long used the stepwise ascent of keys—the so-called "pump-up modulation"—as a standard practice with the standard effect of intensification and energy-gain.[45] But film music's use of stepwise key relationships is more varied and flexible, and can take on a wide variety of meanings, depending on the narrative.[46]

Another concept adapted from the context of Wagner operas (Bailey 1969, 1977), **directional tonality** entails beginning and ending a work in different keys.[47] Scholars who study directional tonality in nineteenth-century music agree that it is used most often in texted music (i.e., songs and operas) and that it is typically used to reflect dramatic elements in the text.[48] The narrative-centric nature of film thus makes it a natural setting for directional tonality. If the beginning and ending keys are used associatively in a film, meaning can be interpreted about the move from one key to the other. Often, directional tonality involves the tonal envelope of a soundtrack moving up or down by step, implying some form of progress or devolution, based on the associations we make with upward and downward motion (as mentioned above). Unlike the stepwise relationship described in the previous paragraph, however, the keys involved in directional tonality are not pitted against one another or coupled in any salient way throughout the film; it's simply a matter of the arc formed between the beginning and ending keys.

Another form of tonal envelope in film is the **tragic-to-triumphant arc,** which illustrates a trajectory of victory and progress when a film begins in a minor key and ends in the parallel major—utilizing the parallel relationship on a larger teleological scale. Robert Hatten first defined "tragic-to-triumphant" as an "expressive genre" in the context of nineteenth-century music, to characterize (for instance) the "triumph of the will" of C major over C minor across the four-movement span of Beethoven's Fifth Symphony.[49] The tragic-to-triumphant arc represents one variety of tonal "happy ending" in film soundtracks.

---

44. Bailey (1977) traces the nineteenth-century rise of such tonal shifts to the ubiquitous use of harmonic sequences, established in earlier centuries. See McCreless (1982 and 1996) for nuanced discussions of stepwise key relations in nineteenth-century music.

45. See Doll (2011) for more on the "pump-up modulation" in rock music (including the charming plethora of inventive nicknames devised by other scholars). In Doll (2017), he categorizes the "pump-up" as a transformation rather than a modulation per se (204–206).

46. Lehman likewise supports the notion of expanding the expressive range beyond the tension/relaxation binary (2012b: 120–123). For detailed discussions of "expressive tonality" in film music (on a more localized level), see Lehman (2012b: 113–124 and 2018: 49–65).

47. I am, of course, stripping away any caveats of monotonality or third-relations stipulated by some theorists (in the context of nineteenth-century music) as components of directional tonality.

48. See, for example, Krebs (1981) and Stein (1985).

49. Hatten (1991: 76 and 82, respectively). See also Hatten (1994).

Operating at a more localized level, **cadential frustration** involves the interruption of cues on dominant-functioning chords, obstructing tonic resolutions as an analogy for obstruction in the narrative. While cadential frustration is not *itself* key-specific, it can be paired with associative tonality or other tonal components. For instance, systemic disruption of cadences may occur in the context of tonal agency, where thwarted cadences illustrate a character's thwarted intentions (especially in music the characters themselves diegetically produce). Or a character's musical agency may be sabotaged by another character, where cadential closure is obstructed as part of a power play. Frank Lehman (2018: 201) coins the phrase "cadential mickey-mousing" to describe the common film practice of aligning cadential resolution with important narrative/visual events—and cadential frustration can be a potent manipulation of that phenomenon, since viewer-listeners have grown accustomed to this type of harmonic-dramatic alignment. Cutting off cues on "resolution-craving dominants" (Lehman 2018: 202) can sonically convey (at least subconsciously) to the listener a sense of obstruction.

Other elements of **functional tonality** can operate on a larger scale, if we zoom out on the tonal relationships between chords and expand them to relationships between keys. For example, we can think of the *key* of G major as having a dominant function to the *key* of C major. Or alternately, C major could be considered the subdominant of G major—which key we interpret as the "tonic" depends on the narrative context and how keys are being used in a particular film. If C major is the key of denouement in a film, then G major could serve as a dominant springboard into that key; but if G major is the end-goal key in a film, narratively speaking, then C major might provide plagal resolution to that key. At this zoomed-out level, functional relationships between *keys* can convey the same types of conventional implications that go along with the equivalent *chordal* relationships—for instance, the preparatory nature of a V key leading to a I key (or conversely, the open-ended feeling of ending on a V key) and the redemptive nature of a IV key leading to a I key.[50]

On an even broader scale, some film soundtracks allude to an overarching **meta key,** in that the main keys of the film outline scale degrees of one particular key, the tonic (key) of which occupies a significant position ("significant" in terms of narrative importance and/or total amount of airtime). For instance, a film in which D major is the most important key, with the keys of F♯ and A occupying roles of secondary importance, thus implying the tonic-triad members of the D-major meta key. A filmic meta key does not require (or imply) rigid adherence

---

50. To adapt Murphy's adaptation of Lehman's (2018) adaptation of Babbitt (1987), "any analytical accounts of intertriadic relationships should strive to reflect Babbitt's perceptual sort of context dependency" (Murphy 2023 forthcoming) in which "the meaning of a musical event hinges on its musical surroundings" (Lehman 2018: 130). Of course, Murphy, Lehman, and Babbitt are discussing different types of progressions and musical settings, but their points are easily generalized to apply to other contexts.

to all the traditional harmonic structures/relationships within a key, but rather, a general sense of "keyness" (that arises from salient factors such as the ones listed here).

3. Elements that Use Keys Abstractly

This category covers the syntactic rather than semantic use of keys—in other words, keys delineating logistical or structural elements of a film, rather than adding meaning to the narrative. For example, keys may be used to designate beginning/middle/end stages of a trajectory, or correspond to visual or temporal elements of a film, without being actually *associative* or correlating to specific meaning. With **tonal symmetry**, the overarching tonal layout of a soundtrack exhibits symmetry (either reflectional or translational)—that is, the entrance and exit of keys across the span of a film forms a symmetrical pattern. Tonal symmetry often mirrors some sort of symmetry in the narrative or structure of the film, so it provides rich opportunities for interpretation. In chapter 4 we look at in-depth analyses of two films featuring tonal symmetry.

A less profound but more prevalent technique is what I have coined **jump-cut tonality**, the phenomenon of diegetic music abruptly shifting to a different key across a filmic jump cut to humorously convey the passage of time. The tacit implication of a tonal jump cut is that such an inordinate amount of time has passed that the music has now modulated to a new key. The specific keys are not important, nor is the relationship between them; the change of key is all that matters in this context. This kind of tonal joke gets a laugh even from viewer-listeners with no formal knowledge of keys or modulation, who instinctively understand that the tonal disjunction implies the impatience of diegetic listeners over the course of a tediously long musical number.

CLOSING THOUGHTS

The elements described here do not represent an exhaustive list, by any means, and should be considered a starting point for the exploration of tonal relationships in film soundtracks. At the heart of the matter, the analysis of film tonality can be reduced to these core questions: Do keys matter in film or not? How does the particular layout of keys in a soundtrack allow us to think about music (and other pitched sound) with respect to narrative structure? What do we gain from including key as one of the parameters of film music analysis? The three analysis chapters that follow will answer these questions by illustrating the rich productivity of interpreting tonality in film. Chapter 2 presents films that feature the tonal elements described here, while chapters 3 and 4 are themed around particular tonal elements, to show how tonality allows us to connect seemingly unconnected films.

2

Tonal Analysis of the Soundtrack

Now that we have laid the theoretical foundation for interpreting film tonality, let's start analyzing. In this chapter we explore five films that engage the tonal elements presented in chapter 1 (except for directional tonality, meta key, and tonal symmetry, which are presented in the themed analysis chapters, 3 and 4). The films in this chapter are arranged such that each film shares (at least) one key tonal element in common with the next. Pitched sound effects contribute to the tonal design of every one of these films, but this chapter focuses almost exclusively on music, so that the tonal contribution of sound effects can be given a full discussion in chapter 5.

THE ENGLISH PATIENT (1996)

Elements featured: associative tonality, transposition, intertextual tonality, tonal pairing, tonal coda, *Leittonwechsel* relationship.

RELEVANT PLOT SYNOPSIS

During World War II, a nurse named Hana (Juliette Binoche) leaves her unit to care for the "English Patient" (Ralph Fiennes), a critically injured man who cannot remember who he is. The whole film centers around the Patient's struggle to recall his identity, and through numerous flashbacks, he slowly begins piecing together his past . . .

He was a Hungarian cartographer named Almásy, working with an English mapping expedition in the Sahara Desert. Almásy falls in love with Katharine (Kristin Scott Thomas), the wife of an expedition member, and they embark on a passionate love affair.

During the expedition, Almásy is elated to discover a cave containing ancient paintings of swimming figures (showing that there was once a lake in the Sahara Desert). The Cave of Swimmers becomes a magical, special place for him.

When the expedition returns to the Sands Club in Cairo, Katharine begins to pull away, which causes Almásy to jealously and dangerously lash out (drunk) during a public banquet dinner. Katharine's husband (Colin Firth) finds out about her affair and crashes his plane while attempting to hit Almásy in a murder-suicide. Katharine is critically injured in the crash, and Almásy carries her to the Cave of Swimmers. He returns to Cairo and begs British officers to send help, but because of his non-English name, the British assume he is a German spy and imprison him. By the time Almásy escapes and returns to the Cave, Katharine is already dead. After this, Almásy crashes his own plane and suffers severe burns and partial amnesia. Bedouins find him, and this is where the film began.

After recalling his entire past, the Patient asks Hana to administer a lethal dose of morphine. After doing this, Hana leaves to begin her new life, now that the war has ended.

SUB-PLOTS:

A stranger by the name of Caravaggio (Willem Dafoe) joins Hana and her Patient. At first, he suspects the Patient is one of the spies responsible for his (previous) interrogation and torture (during which his thumbs were cut off). But he eventually learns the truth when the Patient recounts his story.

While caring for the Patient, Hana has a sweet love affair with Kip (Naveen Andrews), a bomb-defusal specialist who is sweeping the area for mines. Kip asks Hana to come away with him when his unit is departing, but Hana makes the heartbreaking sacrifice to stay behind, out of a sense of duty (to continue caring for the Patient).

The characteristic quality of *The English Patient* is exotic, hazy mystery. The narrative is unfurled in tantalizing bits and pieces, with a constant stream of flashbacks obscuring the linear progression of events. The various locations visited throughout the narrative are not explicitly identified but are hinted at with intriguing details. The film's opening visuals feature nebulous imagery that could possibly be the sinuous contours of sand dunes or the curves of a female body. Likewise, the film's sound also opens with an aura of mystery, starring three distinct aural layers. First we hear tinkling glass bottles evoking a Middle-Eastern bazaar, followed by an Arabic prayer chanted by a distant male voice, and a third entry by a female voice singing a Hungarian folksong ("Szerelem, Szerelem") (0:00:53).

Contrapuntally, these three separate "voices" intermingle to insinuate a sense of locale that is at once easy to characterize (a desert bazaar setting) and difficult to place (is it Middle Eastern, or North African, or Hungarian?). Harmonically, too, these three voices interact to create an "exotic" tonality to Western ears. The glass bottles clink out the pitches F, G♯, and A, with enough constancy that they create a consistent sonic texture. The male voice enters on these same pitches (F-G♯-A) before establishing D as the tonic, thereafter oscillating mainly between D and A. The female voice sings a more complex melody, which amalgamates with the clinking bottles and the male chanting to suggest a nebulous realm between D natural minor and D "Hungarian minor" (harmonic minor with ♯4̂).

These three musical layers fuse together to create an aural landscape centered around a D tonic, but with its precise tonal identity hazed in a bit of mystery. This tonal mystery is corroborated by the *functions* of the three sound layers: the female-voice layer is clearly intended as nondiegetic (with its foregrounded volume level), but the male-voice and clinking-glass layers have a grainy, backgrounded sound quality that implies they are diegetically emanating from a particular setting. However, this setting is not visually revealed (until far later in the film, when Almásy is walking through a bazaar, and we hear the male chanting and clinking glass sounds in their natural environment), so these two layers take on a detached, acousmatic quality that adds to the deliberately mysterious air of the film's opening sequence. Further discussion of the meaningful use of sound effects in this film takes place in chapter 5.

The "Szerelem, Szerelem" folksong fluctuates between the poles of D and A, and the last verse of the song ends on an A, with the expectation that it will rebound to D (as has occurred in all previous verses). Instead, the next cue (Gabriel Yared's "Desert Theme") dovetails with the previous cue and picks up on that D pitch, to introduce the key of B minor (0:02:34). The music and sound effects of the opening have built up to this introduction of B minor, which is the main key area of the film. Preceding it, the enigmatic, slightly ambiguous intimation of D minor unsettled the listener's sense of (tonal) balance—like sands shifting beneath the feet—before solid footing finally arrives with the key of B minor. With all the meandering movement of the plot, B minor is the tonal locale of Almásy's narrative—the connecting thread that binds together all the varying senses of time, place, and identity (of Count Almásy/"English Patient"). There are a handful of preexisting and original cues (composed by Gabriel Yared) in the key of B minor, and all of them are connected to Almásy. The key of B minor accounts for the majority of musical time in the film, and only occurs when Almásy is part of a scene—never in his absence. So deeply coupled with Almásy is this key that in the one instance in which Yared's "Mortality Theme" is featured in the context of another character (Kip mourning the death of his friend, Sergeant Hardy), the theme is transposed (from its original B minor) to the key of E minor.

Juxtaposed against the mystery and melancholy of B minor, the *Leittonwechsel*-related key of G major represents fulfillment—both romantic and personal. G major occurs in the form of three original Yared themes and six preexisting works. Given that this is a love story between Almásy and Katharine, the most frequent manifestations of G major occur with regard to the romance between these two characters. The *preexisting* works in G major accompany moments of romantic fulfillment that occur in *public* settings (such as Rodgers and Hart's "Where or When" playing the first time Katharine dances with Almásy at the Sands Club [0:37:50], or the party guests singing "We Wish You a Merry Christmas" as

Katharine walks through the crowd to meet Almásy for a sexual tryst [1:19:11]), while the *original* G-major themes accompany *private* moments of romantic fulfillment (such as the consummation of the abovementioned sexual tryst [1:20:25], and the moment Katharine finally tells Almásy she loves him [2:11:10]).

Two of the G-major themes ("Love Theme" and "Passage of Time Theme") are sometimes manipulated to reflect the tempestuous nature of their love affair, and tonic resolution is thwarted through the intrusion of powerfully oppressive nonchord tones. To illustrate G-major *un*fulfillment, the music never actually reaches a proper tonic triad but spends most of its time stuck on ♭II; and where there *should* be tonic resolution, we have an "unfulfilled" tonic, troubled with a heavy ♭2̂ in the bass rather than 1̂ (along with a cloud of other dissonant pitches). This harrowing tonic frustration underscores moments of romantic frustration: when Almásy first touches Katharine but can go no further (during the sandstorm) (1:01:22), when Katharine looks through Almásy's book in a vain effort to get closer to him (1:05:19), when Almásy and Katharine realize they can never have a future together (in the bathtub, after their first sexual encounter) (1:15:01), and when Almásy says that Katharine died because she loved him (2:21:28). During this last incident, Almásy also mentions that Katharine died because he "had the wrong name" (Almásy wasn't able to reach Katharine in time to rescue her because British officers detained him as a German POW, due to his foreign name), and this explains why Almásy has now blocked out his own wretched name, which cost him the woman he loved. At this precise moment, Yared's "Love Theme" ends on its most disturbing chord yet—a minor-major seventh chord (B, D, F♯, A♯)—which then turns into the B minor of the ensuing "Oriental Theme" (see Video 1 ✱).[1] Thus Almásy's G-major unfulfillment leads to the Patient's B-minor loss/suppression of identity.

In stark contrast to the lush, decadent turmoil of Almásy's and Katharine's G major (the "Love Theme" and "Passage of Time Theme," with their almost crushing chromatic load, extravagant orchestration, and effusive dynamic swells), Hana's G major is wholesome, clean, and light. It begins with Hana playing the G-major Aria from Bach's *Goldberg Variations* on a piano (0:43:51), which brings Kip into her life (he sprints across the countryside toward the sound of a piano being played, to search for a booby-trap mine hidden in the instrument), fulfilling her mother's prophecy that she would "summon her husband by playing the piano." The G-major "New Life Theme" (which is directly modeled after the *Goldberg Variations* Aria) further propels their romance (1:49:29), as Hana follows the trail of candles Kip laid out for her, leading to their sexual union. This sparkling theme (with it clean, solo-piano scoring and sparse contrapuntal texture) is

---

1. Incidentally, this disturbing chord is the "Hitchcock Chord" made famous in Bernard Herrmann's 1960 score for *Psycho*, first discussed by Royal Brown (1982) and (1994: 153). I leave it to the reader to muse over the possible intertextual connections here.

reserved exclusively for the pure-hearted Hana and is never used in the context of Almásy or Katharine.

As mentioned above, G major is also used for non-romantic fulfillment, such as the moment Almásy discovers the Cave of Swimmers and bursts into laughter of joy and amazement at such a profound discovery (0:51:32), and Hana's euphoric experience of soaring through the cathedral ceiling to explore the exquisite frescoes of Piero della Francesca (1:51:41).[2] And the film ends with the "New Life Theme" when Hana rides off into the sunset after the war's end (2:36:13), with a smile on her face at her sense of duty fulfilled (having sacrificed her personal happiness to stay behind and take care of Almásy until his death).

"The heart is an organ of fire," Hana reads aloud to Almásy (from his own writing on the inside wrapper of a firecracker), at the moment a church bell in the distance rings out with B♭ major overtones (1:16:49). This moment encapsulates the role of B♭ major in this film, in that every instance of this key accompanies a display of Almásy's fiery passions. The first moment Katharine and Almásy meet, the resplendent B♭-major "Rupert Bear" theme soars as their airplanes fly side-by-side (0:23:54), while Almásy watches her with almost "predatory" focus as their passion first takes flight.[3] Almásy fixates intensely on her once again during their second meeting (at the Sands Club), completely indifferent to the presence of her husband, while the big band plays a jaunty piece in B♭ major (0:36:44). The band plays another B♭-major number while Almásy watches Katharine and prepares to ask her to dance with him (0:37:31).[4]

Much later on, a third big-band tune in B♭ major introduces the scene of the disastrous formal dinner at the Sands Club (1:44:48), in which Almásy is so frenzied with passion that he erupts in an uncontrollable, hot-tempered outburst. The most climactic B♭-major incident takes place while a roomful of Christmas party guests sing "Silent Night" in B♭ major (1:19:54), while Katharine and Almásy engage in a recklessly uninhibited sexual encounter just outside the door. Interestingly, the "Love Theme" in G major is foregrounded over top of "Silent Night," highlighting the fact that Katherine and Almásy are (in that moment) engulfed in their own private world (of G major fulfillment), despite being in the midst of a large crowd of people (separated from the party only by an open doorway). This contrapuntal overlay of B♭ major and G major fuses together the pinnacles of Almásy's passion and fulfillment.

2. In the rafter-mounted rope harness Kip set up for her, to bring some beauty into her war-ravaged life.
3. "Predatory" is the word Katharine later uses to describe Almásy's attention toward her.
4. There are a handful of big-band tunes in this film that are not listed in the credits or soundtrack (or anywhere else on the internet), and they fail to be identified by Shazam, and even the jazz aficionados I have asked are unfamiliar with them. I can only assume these pieces were composed (uncredited) by Gabriel Yared for the film.

D major is the key of Almásy's jealousy. The song "Manhattan" in D major plays during the aforementioned disastrous Sands Club dinner (1:45:08), when an aggressively drunk Almásy lashes out because he is jealous of Katharine's love. (In this scene the "fiery passion" key of B♭ major leads directly to the "jealousy" key of D major.) Almásy's second jealous conniption is also accompanied by a big-band tune in D major (1:47:27), when he furiously berates Katharine for dancing with another man and accuses her of sleeping with other men.

The soundtrack for *The English Patient* is based almost entirely around the twin pillars of B minor and G major, with almost three-quarters of the film's total musical time allocated to these two keys.[5] The *Leittonwechsel* relationship between these keys has significant implications in the narrative, which also tie in to a broader trend in modern film-score practice. As discussed in the previous chapter, the *Leittonwechsel* transformation has become a stock technique in film music for depicting contemplation over the loss of a loved one, and this sentiment is palpable between the keys of G major and B minor in *The English Patient*. G major celebrates Almásy's and Katharine's romance, and B minor is the key in which Count Almásy/"English Patient" unravels the enigma of his identity. Since most of the narrative is presented to us through his reminiscences, the *Leittonwechsel*'s contemplative sense of loss (of a loved one) reflects *his* feelings, as he recounts the passion of his love for Katharine and the pain of her death. This relationship is especially meaningful because Almásy's inability to recall his own identity is directly linked to his contemplation of Katharine's loss—he says that Katharine died because he "had the wrong name" (as discussed earlier), thus his self-inflicted amnesia results from his contemplation of the loss of their love. Reducing that latter phrase to a (conceptual) musical equation, we get the following:

self-inflicted "amnesia" results from contemplation of loss of love
B minor = *Leittonwechsel* × G major

Viewing the tonal relationship from this perspective implies that B minor arises from a transformation to G major—that the key of G major "came first" in the chicken-or-the-egg cycle of tonal reciprocity.[6] As it turns out, the primacy of G major can be corroborated in both the context of the narrative (Almásy had to experience love *before* he could punish himself for losing it) *and* the context of the soundtrack. The key of G major in this film is an intertextual homage to the Aria of Bach's *Goldberg Variations*, which was selected as an anchor for the soundtrack early on and also served as the compositional model for the important "New Life

---

5. Approximately 44 percent in B minor and 29 percent in G major.
6. For an in-depth discussion of tonal reciprocity, see Cohn (2012: 43–58).

Theme" (written in G major).⁷ The *Goldberg* Aria is featured prominently in the film as diegetic music (played by Hana during a pivotal scene), and Yared composed the "New Life Theme" as its nondiegetic counterpart—resembling the Bach in both character and key. Thus the choice of B minor for the key of "lost identity" is based on its *Leittonwechsel* relationship to G major.

These two main key areas of the film are employed in a closing-credits tonal coda. The narrative ends in G major ("New Life Theme," overlaid with a separate layer of Hungarian improvisational singing), and this is where the closing credits begin. After two G-major cues ("Love Theme" and "Passage of Time Theme"), the music settles back to B minor ("Desert Theme," "Mortality Theme," and "Én Csak Azt Csodálom"), the original (and main) key of the narrative. Thus the major tonal poles of the soundtrack are represented in miniature form during the closing credits. (We might even go so far as to interpret the ordering of keys in the tonal coda as a confirmation of the aforementioned *Leittonwechsel* causality, by beginning in G major and ending in B minor.)

MOONRISE KINGDOM (2012)

Elements featured: associative tonality, tonal wink, intertextual tonality, tonal pairing, parallel relationship, relative relationship, tragic-to-triumphant arc.

RELEVANT PLOT SYNOPSIS

Twelve-year-old orphan Sam (Jared Gilman) is attending Camp Ivanhoe, a Khaki Scout summer camp led by Scoutmaster Ward (Edward Norton). Dark and troubled Suzy (Kara Hayward), also twelve, lives on the island with her family, whom she hates. Sam and Suzy, both introverted, intelligent, and mature for their age, met the previous summer at a church performance of Britten's *Noye's Fludde*, and have been intimate pen pals ever since. Having fallen in love over the course of their correspondence, they have made a secret pact to run away together. They hike, camp, and fish together in the wilderness with the goal of reaching a secluded cove on the island. They are pursued by the local policeman Captain Sharp (Bruce Willis), Suzy's parents, Scoutmaster Ward, and his Khaki Scout troop. They eventually find Sam and Suzy, and Suzy's parents take her home. Captain Sharp contacts Sam's foster parents only to learn that they no longer wish to keep him. Sam stays with Captain Sharp while they await the arrival of Social Services (Tilda Swinton), who plans to place Sam in a Dickensian orphanage and reform him with electroshock therapy.

A faction of the Khaki Scouts has a change of heart and decide to help Sam and Suzy. They enlist the efforts of another nearby Scout camp (Fort Lebanon) to help

---

7. Laing (2003: 89ff.) discusses director Minghella's early selection of the Bach for this soundtrack and composer Yared's subsequent work building the "New Life Theme" around it.

TABLE 1 Britten works excerpted in *Moonrise Kingdom*

Britten Work	Movements	Timestamp
Young Person's Guide to the Orchestra	Themes A–F Theme M Fugue	0:01:58
Simple Symphony	II. Playful Pizzicato	0:04:41 0:13:12 0:14:18
Noye's Fludde	5. "Noye, Noye, Take Thou Thy Company" 9. "Noye Take Thy Wife Anone" 11. "The Spacious Firmament on High"	0:17:18 1:20:20 1:17:37 1:23:52
Midsummer's Night Dream	Act II: "On the Ground, Sleep Sound"	0:31:49 0:46:28 0:47:41
Friday Afternoons	3. "Cuckoo!" 12. "Old Abram Brown"	0:49:09 1:25:34 1:00:09

Sam and Suzy escape and form a new life for themselves. Before leaving, Sam and Suzy are married in an unofficial "wedding" ceremony. Sam and Suzy never make it to their destination and are instead pursued by the aforementioned search parties. A violent storm and flash flood strike, and Captain Sharp apprehends Sam and Suzy on the steeple of the church (in which they first met during *last* year's performance of *Noye's Fludde*), during *this* year's performance of *Noye's Fludde*. The steeple is destroyed by lightning, but everyone survives. At the height of the storm, Captain Sharp decides to adopt Sam, thus saving him from the orphanage and allowing him to stay on the island, where he can maintain his friendship with Suzy.

If Bach's *Goldberg Variations* provided the tonal backbone for *The English Patient* soundtrack, the music of Benjamin Britten serves as the entire skeletal structure of *Moonrise Kingdom*. Not only is Britten's music featured prominently in the soundtrack, but it is also central to the overall filmic narrative. The film's story (not the *plot*) begins at a children's summer camp performance of Britten's *Noye's Fludde* (where Sam first meets Suzy) and ends with a performance of the same (one year later). Sam's story loosely parallels the "Noah's Flood" narrative, in which a climactic flood leads to salvation. On a musical level, excerpts of Britten's works (Table 1) are widely used and integral to the film's soundtrack. In particular, the film utilizes Britten works that are geared toward children, to evoke the sense of childhood, innocence, and whimsy that characterizes *Moonrise Kingdom's* child-centric world.

Britten's music also shapes the original scoring in this film. Alexandre Desplat's seven-part "The Heroic Weather-Conditions of the Universe" suite in G minor is an homage to Britten's *The Young Person's Guide to the Orchestra*, complete with an elucidatory narration of the work's instrumentation in Part 7 (narrated by Sam, during the closing credits). All seven variations are in G minor, which is the key that accompanies Sam's journey in search of love and acceptance. The first hint of G coincides with the first hint of Sam: in a deliciously subtle detail so characteristic of Wes Anderson's filmmaking style, Sam's scout registration card is shown with the box "G" checked (Figure 4), while a phone rings on the pitch G as the camera zooms in for a close-up of Sam's picture (0:11:24).[8]

This tonal wink at G corresponds with our first visual introduction to Sam, and when he officially begins his journey (with Suzy by his side), the key of G minor begins also (with Part 1 of "The Heroic Weather-Conditions of the Universe") (0:21:14). G minor underscores Sam's entire expedition, and sees its last iteration (within narrative space) when the storm's climactic flood brings Sam's journey to an end. Then, just as a sound effect *commenced* the key of G, another sound effect *concludes* the key when a bell tolls on G as Sam thanks Suzy for marrying him and acknowledges that their time together has likely reached its end (1:22:38). G minor now gives way to the parallel-related G major (with "The Spacious Firmament on High" from Britten's *Noye's Fludde*), when the flood brings Sam the salvation he was seeking: a place where he *belongs* (this G-major music marks the moment Captain Sharp states his intention of adopting Sam) (1:23:52). In Britten's opera this G-major chorus comes after Noah is delivered from the flood; in Anderson's film this G-major chorus marks Sam's *literal* deliverance from the flood (caused by the Black Beacon Storm) and *figurative* deliverance from the *Sturm und Drang* of his G-minor angst (finding peace in the loving care of Captain Sharp). The key of G major in this film seems to spring forth (intertextually) from this pivotal, narrative-driven Britten selection. And the large-scale movement from G minor to G major outlines Sam's tragic-to-triumphant arc like a rainbow at the end of the storm.

Like G major, the key of B♭ major also derives intertextually from Britten's *Noye's Fludde*. The conspicuous B♭-major bugle-call motif present in all three of the *Noye's Fludde* movements is echoed by a similar B♭-major bugle call in the "Khaki Scout Marches" theme by Mark Mothersbaugh. Whether in the form of a bugle call or an operatic chorus, B♭ major heralds new beginnings in this film. It underscores Sam and Suzy's momentous first meeting (at last year's summer camp performance) *and* their momentous second meeting (when they meet to run away together) (0:17:18), it introduces Camp Ivanhoe (0:05:29) and later Fort Lebanon (1:05:53), it signals the move to higher ground after the flood has ensued (1:14:58), it accompanies Captain Sharp as he rescues Sam from himself (during the storm) (1:20:20), and it marks Scout Master Ward's fresh start in rebuilding the ravaged Camp Ivanhoe (as well as his ravaged career) (1:24:20).

---

8. The phone ring's main pitch is G, with an E♭ resonating below it.

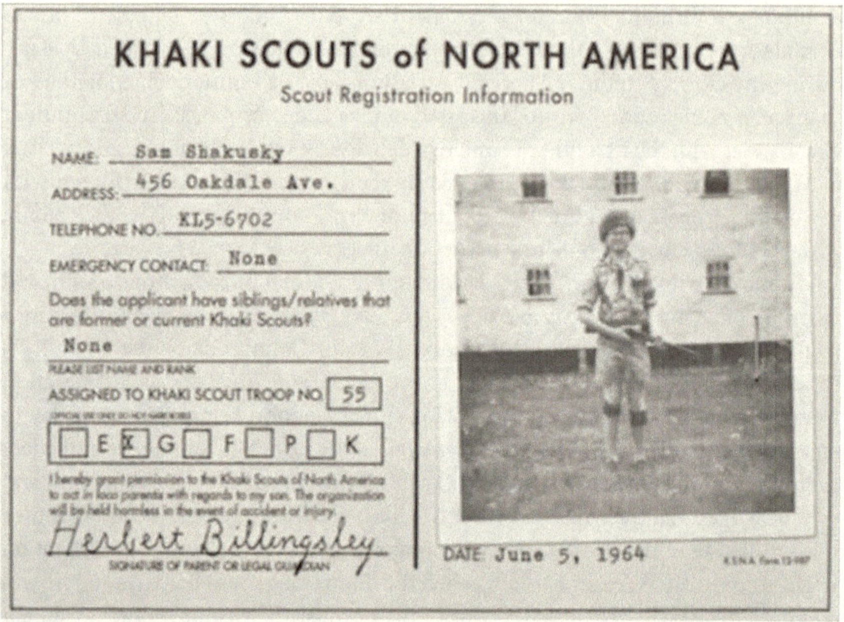
KHAKI SCOUTS of NORTH AMERICA

Scout Registration Information

NAME: Sam Shakusky

ADDRESS: 456 Oakdale Ave.

TELEPHONE NO. KL5-6702

EMERGENCY CONTACT: None

Does the applicant have siblings/relatives that are former or current Khaki Scouts?

None

PLEASE LIST NAME AND RANK

ASSIGNED TO KHAKI SCOUT TROOP NO. 55

☐ E ☒ G ☐ F ☐ P ☐ K

I hereby grant permission to the Khaki Scouts of North America to act in loco parentis with regards to my son. The organization will be held harmless in the event of accident or injury.

Herbert Billingsley

SIGNATURE OF PARENT OR LEGAL GUARDIAN

DATE June 5, 1964

FIGURE 4. Sam's Khaki Scout registration card, with the "OFFICIAL USE ONLY" box "G" checked.

The tonal design in this film is constructed around two main keys (G major and B♭ major), both of which are derived intertextually from preexisting works that figure largely in the narrative. Sam's story strongly parallels the story of *Noye's Fludde*, and this parallel is reflected in the tonality as well as the narrative. And both G major and B♭ major are closely related to G minor (parallel and relative relationships, respectively), the key Desplat chose for his original seven-part suite that underscores Sam's journey.

BREAKING AND ENTERING (2006)

Elements featured: intertextual tonality, tonal coda, stepwise relationship.

RELEVANT PLOT SYNOPSIS

Will (Jude Law) shares an unsatisfying life with his life-partner Liv (Robin Wright) and her troubled daughter, both of whom are emotionally distant and continually push Will away (despite his desire to get closer). Through a chance encounter, he meets a refugee named Amira (Juliette Binoche) and her son, both of whom *need* Will (in their different, subtle ways). Feeling attracted to Amira's warmth and vulnerability, Will finds excuses to visit her home, and they begin a romantic affair. They both have conflicted feelings about the affair, and both feel trapped between two incompatible worlds (Will is trapped between his love for his partner and his paramour, and Amira is trapped between her roles as a mistress and a mother). Their discordance is further complicated by the fact that Amira's son and his criminal friends repeatedly burglarize Will's architecture firm; Will knows it, and Amira knows, but neither one knows *the other* knows. If Will reports the crime, Amira and her son will be deported. They each struggle with how to deal with this situation, while trying to keep the other from finding out.

As in *The English Patient* and *Moonrise Kingdom*, intertextual tonality plays an important role in this film—but rather than the lovelorn *Leittonwechsel* relationship of the former, or the redeeming parallel relationship of the latter, the *Breaking and Entering* soundtrack features keys engaged in a conflicting stepwise relationship. This Anthony Minghella film is centered around the tension between the two worlds of protagonist Will: his cold Swedish partner Liv with her disturbed daughter and his warm Serbian paramour Amira with her disturbed son. Will is torn between his love and sense of duty toward both of these women and their fragile children. Tonally, this tension is played out with stepwise relationships throughout the film. Not only do the original works (composed by Gabriel Yared in collaboration with the British electronic band Underworld) in this film make heavy motivic use of whole-steps (both melodically and harmonically), but many of them also oscillate between whole-step related tonics. For example, the track entitled "Will and Amira," which fluctuates between C minor and D minor tonics—sometimes appearing in one key, sometimes appearing in the other, and sometimes modulating back and forth between the two during a single cue. The music modulates from one tonic to another when the scene transitions from one of Will's worlds to the other. The juxtapositions from one cue to the next are very often whole-step related as well. The keys of A minor and B minor are paired together in a prominent stepwise-related tension pair, as are C minor and D minor.

The key of C minor in this film emanates from Bach's Sinfonia No. 2 in C minor (BWV 788), which facilitates a crucial montage sequence in the middle of the film (1:03:21). Immediately preceding it, the "Will and Amira" cue fluctuates from D minor to C minor to D minor (1:01:49); then, in lieu of another oscillation to C minor, the Bach Sinfonia picks up the key of C minor. The Bach concludes with a Picardy-third resolution in C major right when both women (Liv and Amira) are concurrently (but separately) discussing Will's happiness, in cross-cutting scenes during the montage. But this brief ray of C-major happiness is *immediately* roiled in stepwise conflict, as our discussion of sound effects in chapter 5 shows. Even the music for the closing credits (1:53:54), "Sé Lest" by Sigur Rós, fluctuates continually between two stepwise-related keys (G major and F major). This forms a tonal coda *of sorts* in that the closing credits recapitulate not the specific keys themselves but the tonal motif of whole-tone oscillation explored throughout the film.

THE DARJEELING LIMITED (2007)

Elements featured: associative tonality, tonal agency, singular key, stepwise relationship.

RELEVANT PLOT SYNOPSIS

Brothers Peter (Adrien Brody), Francis (Owen Wilson), and Jack (Jason Schwartzman) join one another on "The Darjeeling Limited" (train) to travel across India. They have not seen each other since their father's funeral a year before. Francis planned this journey to attempt a reunion with their estranged mother (Anjelica Huston), but tells his brothers they are making the journey for spiritual self-discovery. The three brothers are constantly mistrustful of one another, and each is hiding secrets from the others (Jack is hiding his obsession with his ex-girlfriend, Peter is hiding his wife's pregnancy and their marital problems, and Francis is hiding his recent suicide attempt as well as the true purpose of the journey). Francis has created a strict itinerary for the trip and is domineering and controlling of his brothers. Jack begins a secret sexual relationship with the train stewardess Rita (Amaara Karan), despite her relationship with the Chief Steward (Waris Ahluwalia). Peter angers Francis by taking many of their late father's possessions for himself, claiming he was their father's favorite. Between the three brothers, they lug almost twenty pieces of large, bulky, monogrammed luggage belonging to their late father.

Francis eventually reveals that they are on the way to see their mother, who is living as a nun in a Himalayan convent. Peter and Jack are upset upon hearing this, as they still feel hurt over their mother's abandonment. Steadily mounting tensions lead the three brothers to get into a raucous fight, which gets them kicked off the train. Francis reveals a letter from their mother, in which she says she does *not* want them to come. The brothers decide to leave India, go their separate ways, and never return. On their way to the airport, the brothers see three young boys fall into a river while attempting to raft across it. Jack and Francis rescue two of the boys, but Peter fails to save the third, who dies. They take the boys to their rural village and attend the funeral, which reminds them of their father's funeral . . .

(. . . In this flashback the three brothers are on their way to their father's funeral. They stop to pick up their father's Porsche from the repair shop but are incensed to find the car is not ready, after months in the shop. In a cathartic emotional release they seize the Porsche anyway.)

After the village boy's funeral, the brothers take a bus to the airport. Just as they approach the plane, they suddenly rip up their tickets and decide to go visit their mother after all. The reunion is emotional but conflicted. Upon waking the next morning, the brothers find that their mother has abandoned them once again.

The three brothers now prepare to leave India, having attained a new level of peace within themselves and harmony with one another. They have resolved the problems in their personal lives (Jack has decided to move past his toxic ex-girlfriend, Peter has reconciled over the phone with his pregnant wife, and Francis has admitted that his "accident" was a suicide attempt), and they have come to terms with their troubled relationships, with their parents, and with each other. They arrive

TABLE 2 C♯-major and D-major cues in *The Darjeeling Limited*

C♯-major Cue	Type	Timestamp
Indraadip Dasgupta title theme from *Teen Kanya*	preexisting	0:03:56
Ravi Shankar theme from *Pather Panchali*	preexisting	0:07:05
Satyajit Ray "Charu's Theme" from *Charulata*	preexisting	0:16:53
		0:43:06
Ravi Shankar theme from *Apur Sansar*	preexisting	0:19:45
Satyajit Ray "Montage" from *Baksa Badal*	preexisting	0:27:51
Jodphur Sikh Temple Congregation Prayer (Traditional)	preexisting	0:32:42
Satyajit Ray "Mrinmoyee Sad Samapti" from *Teen Kanya*	preexisting	0:35:30
Debussy *Clair de Lune*[a]	preexisting	0:45:14
Amit Trivedi "Typewriter, Tip, Tip, Tip" from *Bombay Talkies*	preexisting	0:54:12
Ustad Vilayat Khan "Arrival In Benares" from *The Guru*	preexisting	1:21:06

D-major Cue	Type	Timestamp
Ustad Ali Akbar Khan title theme from *The Householder*	preexisting	0:05:34
		0:13:35
Satyajit Ray "Ruku's Room" from *Joi Baba Felunath*	preexisting	0:16:03
Satyajit Ray "Manjula's Procession" from *Shakespeare Wallah*	preexisting	0:18:09
Ustad Ali Akbar Khan "Farewell To Earnest" from *The Householder*	preexisting	0:33:19
		1:12:17
Satyajit Ray "The Deserted Ballroom" from *Shakespeare Wallah*	preexisting	0:37:32

NOTE: With exception of the Debussy, all these cues are themes from Indian cinema, mainly films by the legendary filmmaker and composer Satyajit Ray.

[a] The Debussy is written in D♭ major, but the enharmonic (D♭/C♯) distinction is collapsed in our aural perception of film tonality, as discussed in chapter 1.

at the train station to find the train already pulling away. Realizing they can never hope to catch the train if they hold on to their father's bulky luggage, they gleefully discard all his baggage as they run after the departing train. Once onboard, they look back at the abandoned luggage and feel a great sense of relief. They set forward on their journey with a strong new bond of brotherly love.

Like *Breaking and Entering*, the main tonal trope of *The Darjeeling Limited* is the stepwise relationship. Half-step and whole-step relations express the tension of liminal spaces in which the characters find themselves trapped. The tonality of this film fluctuates mainly between C♯ and D major, with the majority of the film's music belonging to these two keys (Table 2).[9]

Instead of an associative distinction attached to either key, tonal meaning arises from the stepwise relationship between them, as successive cues tend to alternate

9. Most of these works are in major mode; there are occasional minor-mode inflections, but for the sake of simplicity, I refer to them all as major mode. In any case, mode is of no particular importance to the tonal design of this film.

FIGURE 5. Stepwise fluctuation of key areas in the first two-thirds of *The Darjeeling Limited*. For visual clarity, keys *not* stepwise-related to their surroundings are designated with unstemmed noteheads. Contiguous stepwise relations end after the key of A major has been attained, two-thirds of the way through the film.

between the two keys. This tonal interplay underscores the on-again, off-again relationships between the characters, which rock repeatedly back and forth like the rhythmic swaying of the train cars they occupy. Among the three brothers, tensions continually fluctuate as permutated sets of two brothers pair off to gossip about the temporarily absent brother. They all struggle in a liminal state between trust and mistrust, as they confide in one another with the proviso "Don't tell Peter/Jack/Francis," but immediately betray one another's confidence the moment they are alone with the other brother. Their relationships with their father and their mother are similarly fraught with emotional pole-reversals, as are the relationships between Jack and Rita, and Rita and the Chief Steward (her boyfriend). The conflicted nature of these relationships is reflected by the dissonant stepwise fluctuation between key areas of contiguous cues (Figure 5). As this tonal staff shows, almost all cues throughout the first two-thirds of the film (the reason for this is discussed below) are related by whole- or half-step to the keys that precede and follow them. Consequently, the tonality of this film is permeated by the pervasive conflict of the dissonant second. Use of the stepwise conflict in sound effects—including their role in the film's ending—is discussed in chapter 5.

Stepwise relations (between the keys of contiguous cues) progressively decrease as the film proceeds, as the brothers stabilize their states of mind and repair their relationships. The incessant stepwise clashing between cues subsides altogether after the key of A major is attained. A major is the key in which the brothers come to terms with their father (their feelings about him as well as his passing). One of the major sources of tension in the brothers' relationships with one another stems from lifelong competition for their father's love. And on a personal level, all three of them have been psychologically stranded between resentment and regret (toward their father) since his death, which has arrested the progress of their personal lives.[10] Once they come to terms with their father-relationships, their brother-relationships finally begin to resolve. And re-bonding as brothers helps them to overcome their crippling awe of their father, and move forward with their lives as adults.

---

10. Francis attempted suicide, Jack is trapped in an obsessive and unhealthy relationship, and Peter is waiting to be divorced by his pregnant wife.

All of this comes to fruition under the aural auspices of A major. Accompanied by the A-major first and fourth movements of Beethoven's Symphony No. 7, the brothers go through a series of steps to liberate their father's Porsche (from the mechanic holding it hostage), which gives them a sense of agency and control (beginning at 0:56:48). At the end of the film, they attain their freedom by finally letting go of their father's belongings, accompanied by the A-major song "Powerman" by The Kinks (1:23:38). "Powerman" begins with v and IV chords, and the A-major tonic is established precisely at the moment the brothers decide to shed the weight of their father's baggage (✶). They finally rid themselves of the burden of their father's approval/disapproval, which allows them to progress as adults—and progress literally too, as they are able to catch their train by discarding his heavy luggage. During this crowning moment, the A-major music is triumphantly foregrounded, drowning out all other sound, along with the dramatic power of slow-motion visuals.

The Darjeeling Limited is the only Wes Anderson film (as of 2022) to feature a soundtrack with *no* original compositions. As such, the tonal design in this film relies entirely on the careful selection and placement of preexisting music and sound effects.

AMADEUS (1984)

Elements featured: associative tonality, tonal agency, parallel relationship, cadential frustration, functional tonality.

RELEVANT PLOT SYNOPSIS

The film opens with Salieri (F. Murray Abraham) as an old man, attempting suicide because he is racked with guilt over having "killed" Mozart. The following morning, Father Vogler (Richard Frank) comes to visit Salieri in the insane asylum and bids him give confession. Salieri tells the priest the story of his life, career, and experience with Mozart (Tom Hulce).

As a boy, Salieri was envious that young Mozart's father, Leopold (Roy Dotrice), devotedly promoted his son's music, while Salieri's *own* father obstructed his musical ambitions. Young Salieri made a pact with God that He might clear the path for his musical career in exchange for his lifelong devotion and service. Salieri's father choked to death in the very next scene, and young Salieri expressed his gratitude to God.

Salieri was appointed as the court composer for Emperor Joseph of Vienna and enjoyed great musical success. When Mozart came to Vienna, Salieri was disgusted to find Mozart immature, unprincipled, and sublimely talented. He began to harbor resentment toward God for favoring a lewd, puerile creature with such inordinate gifts. Mozart found Salieri's music to be completely unremarkable and mocked Salieri's mediocrity at every opportunity. Salieri enlisted his rank and his cronies to hinder Mozart's musical success in Vienna, to little avail. Mozart triumphed effortlessly, and Salieri's resentment (toward God and "His Son") increased.

When Mozart's father died, Salieri seized the opportunity to seek his revenge. Playing on Mozart's feelings of guilt (over having neglected his father), Salieri disguised himself in Leopold's characteristic costume and showed up at Mozart's door to commission a requiem for "a man who deserved a requiem mass and never got one." Feeling confronted by his father's ghost, Mozart's mental state began to falter, as did his health, finances, and marriage. When Mozart collapsed at a performance, Salieri took him home and urged him to finish composing the *Requiem* (which Salieri was planning to pass off as his *own* composition, after Mozart's death). After working on it together through the night, Mozart died in the morning and was buried anonymously in a pauper's grave.

Back in present day, Father Vogler is horrified at Salieri's casual account of his shocking actions. Salieri, however, feels relieved and leaves the room in high spirits, offering his absolution to the other asylum inmates as their "patron saint of mediocrity."

Like *The Darjeeling Limited*, the soundtrack for *Amadeus* consists entirely of preexisting works and derives most of its tonal significance from meaningful harmonic relationships rather than one-to-one correlations between key and meaning. To begin with, a parallel relationship contrasts the dazzling success of Mozart's boyhood (F major) with the frustrating difficulty of Salieri's boyhood (F minor) in a fairly obvious way. Young Mozart performs his Klavierstück K. 33B in F major for the Pope, with his supportive father by his side (0:10:14).[11] In contrast, young Salieri is trapped beside his own father, listening to the "Quando Corpus Morietur" of Pergolesi's *Stabat Mater* in F minor, performed by the church choir (0:11:20). The disparities between their childhoods are conveyed through activity (Mozart actively *produces* music, while Salieri passively and mutely *observes* it), parental involvement (Mozart's father promotes his musical career, while Salieri's father hinders his), status (Mozart is metropolitan, cultured, and wealthy, while Salieri is provincial, plebian, and poor), and mode (Mozart's boyhood is in F major, while Salieri's boyhood is in F minor). The F major of Mozart's childhood is brought on by Salieri's reminiscence that Mozart was his "idol," thus idealizing the key of F major. When he tells the story of his *own* childhood, he deflates the key to F minor, to emphasize the disparity between their boyhood experiences.

Once both characters are grown men, Salieri is subtly portrayed as "simple" through his frequent use of C major—the most basic key, with no flats or sharps in the key signature, which was considered "the key of the neophyte and of the amateur" during the Classical era (Bribitzer-Stull 2006a: 170). And Mozart later aggressively appropriates this key (in a show of tonal agency) to belittle Salieri. The simple key of C major originally belongs to Salieri, beginning with its first instances in the film. We first hear C major in one of Salieri's elementary compositions during Emperor Joseph's keyboard lesson (0:13:14) and then again when Salieri composes an unsophisticated welcome march in C major to celebrate Mozart's arrival at the Viennese

---

11. All timings for *Amadeus* reflect the original two-disc theatrical version, not the director's cut (which now seems to have become the default version). Timings from the second disc are specified as such.

court (0:26:02). As Salieri proudly works hard to compose what he considers to be a noble march, it is reduced to risibility as it becomes nondiegetic accompaniment for Mozart's frivolous wig-shopping montage (0:26:33). Soon after, when this C-major march is performed at court for Mozart's formal entry (0:28:26), Mozart openly ridicules it and then improves it by improvising far superior C-major variations on the theme. (What Mozart improvises in this scene is actually the main A-section theme of his "Non più andrai" aria from *The Marriage of Figaro*. So the scene implies that Mozart not only derided Salieri's little theme, embarrassing him in front of his employer and colleagues, but went on to appropriate it for his own advantage.)

In the first instance, Mozart denigrates Salieri's C-major music by using it to facilitate his silly vanity. In the second instance, Mozart uses it to juxtapose his own superior skills against Salieri's meager abilities. With his first court-commissioned opera, Mozart appropriates the key of C major away from Salieri entirely, using it as the key for his *Abduction from the Seraglio* (of which all the selections featured in the film are in C major). Mozart then uses this work—and key—to steal Madame Cavalieri away from Salieri. Salieri is in love with her and pines for her during their voice lesson together. But despite his efforts to divert her away from Mozart, Madame Cavalieri ends up as the lead soprano in *Abduction from the Seraglio*, singing coquettishly at Mozart in C major while exhibiting body language that clearly implies their romantic involvement (0:39:14). Later, at a masquerade ball, Mozart viciously lampoons Salieri's compositional style with a derisive C-major impression (1:07:18), making Salieri the butt of the ball (Figure 6). Mozart replaces the final C-major tonic chord of this parody with a virtuosic fart, thus conspicuously correlating C major with his scorn for Salieri's simplicity (more on this fart later).

Salieri does attempt to reappropriate the key of C major with his new opera *Auxor* (1:34:08), but Mozart mocks him to his face at the performance, and Salieri's brief moment of tonal agency and triumph (at receiving the Emperor's declaration of "best opera yet written") is rendered bitter to him.

On several occasions the key of G major acts as a sort of dominant springboard to the key of C major, creating instances of inter-cue functional tonality. Leading up to Mozart's abduction of Madame Cavalieri with his C-major *Abduction from the Seraglio*, she sings a G-major scale during her voice lesson with Salieri (0:39:03), which turns into the V^7 chord of Mozart's "Martern aller Arten" aria, facilitating her move away from Salieri and into Mozart's opera (and arms). Later, in trying to convince the court to let him stage *The Marriage of Figaro*, two instances of G major build up to Mozart's final C-major triumph (when he finally succeeds in persuading the court).[12] Mozart begins composing the opera with a G-major

---

12. This victory is especially infuriating for Salieri because he had attempted to use his influence at court to obstruct *The Marriage of Figaro* and had hoped that Mozart would lose favor with the Emperor as a result of his use of banned elements (the libretto and the ballet).

FIGURE 6. Mozart at the masquerade ball, parodying Salieri's compositional style in C major, ending with a great fart in place of the final C-major tonic chord.

selection ("Ah Tutti Contenti") (1:08:45), then uses another G-major number ("Cinque, dieci, venti, trenta") when he beseeches the Emperor to allow the opera to proceed (1:22:48). All of this G-major preparation leads to the victorious dress rehearsal and performance, which both feature the C-major "Ecco la Marcia" (1:24:10). After *Figaro* flops at court, Salieri describes the opera's failure as "a miracle" while accompanied by the G-major "Ah Tutti Contenti" (1:30:12). With this G-major selection, Salieri uses Mozart's failure to vault into his *own* triumphant C-major opera *Auxor* in the very next scene.[13]

The key of D minor plays the dark role of leading Mozart to his death—surely not a coincidental tonal choice, since D minor is the key of Mozart's *Requiem*.

---

13. Publicly, Salieri does win this battle (receiving the Emperor's highest accolade: "the best opera yet written"), even though Mozart ends up spoiling it for him.

Throughout the film numerous pivotal instances of D minor adumbrate Mozart's path to demise. The film opens with the Overture to *Don Giovanni* (0:00:23), as Salieri begins his confession of how he "killed" Mozart. This Overture is played again when Leopold arrives at Mozart's apartment, setting up the father-son guilt complex that Salieri later uses to achieve Mozart's death (0:59:26).[14] Mozart's sense of guilt intensifies greatly after his father's death, and as the Commendatore scene from *Don Giovanni* explicitly expresses a scenario of father-son guilt, Salieri hatches his plot to kill Mozart. The first movement of Mozart's Piano Concerto No. 20 plays as Salieri acquires the black mask Leopold had worn (disc 2, 0:00:02), and the *Don Giovanni* Overture plays when Mozart is stunned out of his senses to find the masked stranger at his door (disc 2, 0:01:38). The *Requiem* "Introitus" accompanies Mozart as he accepts the masked stranger's commission to compose a requiem for "a man who deserved a requiem mass, and never got one" (disc 2, 0:02:37). The *Requiem* "Lacrymosa" plays as Salieri fantasizes Mozart's funeral (disc 2, 0:04:17), while the "Dies Irae" shows Mozart working frantically to compose the *Requiem* (disc 2, 0:12:02). The *Don Giovanni* Overture abruptly interrupts Mozart's moment of disrespectful rebellion (in which he mocks his father's portrait while dancing erratically to *The Magic Flute* Overture in E♭ major), sternly shocking him back into his D-minor state of awe-filled, fatherly fear and guilt (disc 2, 0:17:59) (see Video 2 ✱). And when Mozart does finally die, the "Lacrymosa" from his own *Requiem* marks the moment of passing (disc 2, 0:45:40) as well as his forlorn funeral. Throughout the film the D-minor cues punctuate the step-by-step directions Salieri devises to destroy Mozart.

In addition to large-scale harmonic relationships, this film also exhibits tonal design on a localized level, in the form of cadential frustration. Remembering that this entire narrative takes place not in reality but inside Salieri's version of reality (since his narration facilitates the film), everything we witness is mediated to us through *his* perspective. Thus the music editing can be seen as a facet of Salieri's storytelling, reflecting his state of mind and perception of circumstances. In the context of Salieri's narrative, the conspicuous prevalence of music interrupted on the dominant seems to reflect the trope of "unfulfillable longing"—a term Salieri uses early in the film to describe how he feels about Mozart's music.

First, let's look at the few brief moments of cadential fulfillment Salieri *does* experience, early in the film. When Father Vogler arrives to visit him, Salieri plays (on the

---

14. Leopold characteristically dons an imposing black cape. He is wearing it when he arrives in Vienna to disapprove of Mozart's life choices (getting married without his permission and settling in Vienna). He adds a dramatic black mask to the black cape when he disapproves of Mozart's juvenile behavior at the masquerade ball, and when he beseeches him (unsuccessfully) to return with him to Salzburg. This is the last time Mozart sees his father alive, so when Salieri later appears at his door wearing the same black mask and cape, it has the intended effect of making Mozart feel as though his dead father has returned to haunt him.

Salieri offers his prayer . . .	. . . and God fulfills it
Pergolesi *Stabat Mater* - "Quando Corpus Morietur"	Pergolesi *Stabat Mater* - "Amen"
i - V . . .	. . . V - i

FIGURE 7. Young Salieri offers his prayer to God (synchronized with i-V motion in the "Quando Corpus Morietur"), and God fulfills it (synchronized with V-i motion in the "Amen").

fortepiano) an aria from his opera *Auxor*, which takes him (via flashback) to the performance of this opera, during the heyday of his career. This music ends with a satisfying PAC (perfect authentic cadence) and audience applause (1:35:05), which returns the scene to present-day Salieri (via sound lag), basking in the glow of this gratifying memory. Another cadential conquest occurs when Salieri recalls how he overcame his fatherly obstacle to pursue a career in music (see Video 3 ✱). The "Quando Corpus Morietur" from Pergolesi's *Stabat Mater* accompanies Salieri's earnest prayer to God, as he beseeches Him to clear the path for his musical ambitions (in exchange for his lifelong devotion and service). The music is suspended on a V chord (0:12:28) just as Salieri ends his prayer with a solemn "Amen" and picks back up with a V chord in the *Stabat Mater*'s "Amen" when his father chokes to death—ending with a PAC the moment father is dead in his casket. Salieri made his proposition to God with i-V motion, and God fulfilled His end of the bargain with V-i (Figure 7).

After his father's death Salieri achieves his dream of becoming a musician and wins the prestigious post of Viennese court composer. He revels in returning to that happy time in his life via flashback, "correcting the royal sight-reading" (as Emperor Joseph lumbers through one of Salieri's keyboard compositions), and he beams contentedly as a PAC—of his *own* composing, played by the hallowed hands of the Emperor (0:13:34)—returns him to the present (via sound lag). Incidentally, the transition between these two pivotal scenes (from the F-minor Pergolesi to the C-major Salieri) creates another important instance of inter-cue cadential fulfillment. When Pergolesi's "Amen" cadences in F minor, Salieri feels gratitude for the deed God wrought on his behalf (killing his father). This F-minor work is immediately followed by the Salieri keyboard composition in C major (being sight-read by the Emperor), creating minor-plagal motion (iv–I) between the two cues. This

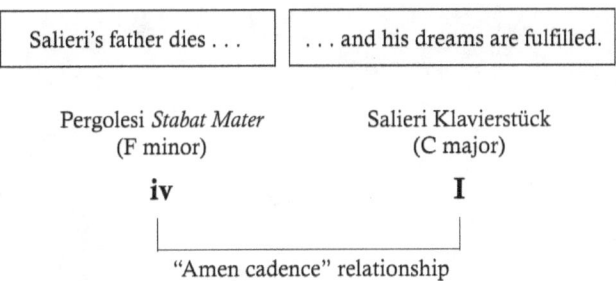

FIGURE 8. Minor-plagal cadential relationship between two adjacent cues, illustrating gratitude and fulfillment across a sequence of events.

plagal motion serves to extend the grateful "Amen" sentiment (the literal text of the Pergolesi) into the next scene—in which young Salieri rejoices in the fulfillment of his dream, being now employed as the Emperor's court composer (Figure 8). Connecting the F-minor death of his father with the C-major fulfillment of his dreams illustrates the sense of causality and righteous fulfillment with which Salieri regards this sequence of life events. It also foreshadows the mighty minor-plagal "Amen" that marks Mozart's death at the end of the film (discussed later).

Cadential fulfillment, however, soon gives way to dominant *un*fulfillment, once Mozart enters Salieri's life. It all starts when Mozart conducts the third movement of his "Serenade for Winds," K. 361 at the Emperor's palace, seizing the spotlight and garnering great applause after the final PAC (0:20:28).[15] Afterward, Salieri gazes at the score and conducts the (metadiegetic) music, and just as he (present-day Salieri, doing a voice-over narration of the flashback) utters the phrase "unfulfillable longing" (0:23:30), the visuals reveal his yearning expression on the particularly poignant V^7 chord shown in Figure 9. In the next moment, after saying he hears "the voice of God" in this music, Mozart swooshes past Salieri and scoops up the score, abruptly aborting the music on another iteration of that same V^7 chord.[16] This poignant V^7 is shattered with the crumpling sound of Mozart snatching the score away from Salieri, illustrating Mozart's power to impede Salieri's ability to achieve cadential closure. Thus, while Mozart conducts this piece to full PAC and triumphant applause, Salieri conducts this piece to painful dominant defeat.

Salieri experiences another instance of cadential castration in the context of his love for Madame Cavalieri. When Salieri coaches Madame Cavalieri to sing scales during her voice lesson (as previously discussed), her G-major scale turns into the

---

15. The film (quite seamlessly) merges the beginning of third movement (Adagio) with the ending of the final (seventh) movement (Rondo), so the final PAC actually belongs to the final movement.

16. The first iteration (see Figure 9) was played by clarinet, while this second iteration is played by basset horn (identical melody, an octave lower).

FIGURE 9. Excerpt (m. 9) from Mozart's Serenade for Winds, K. 361—III. Adagio. The visuals cut to Salieri's look of "unfulfillable longing" precisely at the V⁷ chord on beat 4. The melody (played by Clarinet I) in beat 4 features an accented $\hat{4}$ leaping down to $\hat{7}$, making this an achingly appellative V⁷ chord indeed.

V⁷ chord of Mozart's "Martern aller Arten" aria (from *Abduction from the Seraglio*) in C major, which leads her away from Salieri and directly to her performance in Mozart's opera (see Video 4 ✱). Salieri has that look of "unfulfillable longing" on his face (✱) when she sings this V⁷ harmony, sensing that he is losing her at this very moment to Mozart. Salieri's agonizing inability to consummate cadential closure is especially pronounced when he reviews Mozart's portfolio (0:54:44). Salieri fitfully flips through Mozart's scores, overwhelmed with that "unfulfillable longing" for the "absolute beauty" of Mozart's music, and his agitation increases with each work—each of which is disrupted on the dominant (Table 3).

Notice in Table 3 that even the *progression* of these dominant cutoff chords (if conceptually reduced to a single key and played contiguously) builds up in

TABLE 3 Compositions in Mozart's portfolio, reviewed by Salieri in *Amadeus*

Mozart Work	Ending Harmony
Concerto for Flute and Harp K. 299—II. Andantino	V
Symphony No. 29—I. Allegro moderato	V^6_4
Concerto for Two Pianos K. 365—III. Rondeau. Allegro	V^7
Sinfonia Concertante—I. Allegro maestoso	V^6_4
Mass in C minor, K. 427—I. "Kyrie"	V^7

intensity—$V^5_3\text{-}^6_4\text{-}^7\text{-}^6_4\text{-}^7$—ending with a climactic trill-cadence of the "Kyrie" (Figure 10).[17] This is severed by the jarring sound of pages crashing to the floor, as Salieri drops the entire portfolio in ecstatic agony. This dramatic moment is the peak of Salieri's dominant unfulfillment—climactic in its anticlimactic lack of denouement.

Mozart, however, attains cadential closure with ease and delights in using it to taunt Salieri. He plays the "Vivat Bacchus! Bacchus Lebe!" from his *Abduction from the Seraglio* three times at the masquerade ball, to great cadential effect (1:05:43). The first time (in the style of J. S. Bach), he concludes with PAC, resulting in audience applause. The second time (upside down with hands swapped), he again achieves PAC to great applause. The third time (in the style of *Salieri*), he supplants the PAC with a great fart—*pitched on* $\hat{5}$!—in place of the final tonic chord (see Figure 6), eliciting riotous laughter from the audience (see Video 5 ✶). With this contemptuous cadential obstruction, Mozart mocks Salieri's inability to cadence.

Salieri eventually makes it his life's objective to "hinder and harm" Mozart, using his influential position to put obstacles in Mozart's path. But even when Salieri *does* contrive to halt Mozart in his cadential tracks, Mozart manages to overcome the obstacles and attain closure. In the crusade to stage *The Marriage of Figaro* (as previously discussed), Mozart's "Ecco la Marcia" is suspended on a V chord when Count Orsini-Rosenberg interrupts his rehearsal (1:24:50). A days-long battle ensues over the inclusion of ballet in the opera (forbidden by His Majesty's decree), but it eventually resolves in Mozart's favor when the Emperor attends the dress rehearsal and Mozart piques his curiosity. Capitalizing on this momentary hesitation, "Ecco la Marcia" picks back up on a V chord and cuts directly to the opening-night performance (1:29:20), where it finally resolves to I. Thus the entire battle took place during that prolonged (suspended) dominant chord, and Mozart wins the battle by finally resolving V to I.

In one solitary instance, Salieri does succeed in effectively hindering Mozart's cadential power. After *The Marriage of Figaro* premieres and flops (receiving only nine performances before closing), the "Ah Tutti Contenti" aria accompanies Salieri's

---

17. "Trill-cadences" are typically used in sonata form to close the solo section in preparation for the ritornello return (Hepokoski and Darcy 2006: xxvii).

FIGURE 10. Excerpt (mm. 7–8 after Rehearsal C) from Mozart's Mass in C minor, K. 427—I. "Kyrie." Salieri's look of "unfulfillable longing" as the music cuts off on the V^7 chord of beat 4.

commentary on the opera's failure. His discussion of the ratio of royal yawns to performances (1:32:10) is synchronized with the sequential pattern in this passage (Figure 11), finally ending on a V^7 chord when Mozart enters the discussion (both visually and vocally).[18] Remembering that this reminiscence is crafted by Salieri, he harnesses the descending line of the sequence to symbolically illustrate Mozart's career descending downhill, and Mozart is made to speak on an unfulfilled dominant.

From this point on in the film, Salieri commands the high ground over Mozart. Mozart's health is rapidly deteriorating (thanks, in great part, to Salieri's psychological warfare), and as his life force wanes, Mozart experiences more and more cadential interruption. His rebellious reveling to *The Magic Flute* Overture (as previously discussed) is cut short by crippling guilt on a V chord when the masked

---

18. Salieri: "Three yawns, and the opera would fail the same night. Two yawns, within a week at most. With one yawn, the composer could still get . . ."
Mozart: "Nine performances! Nine—that's all it's had!"

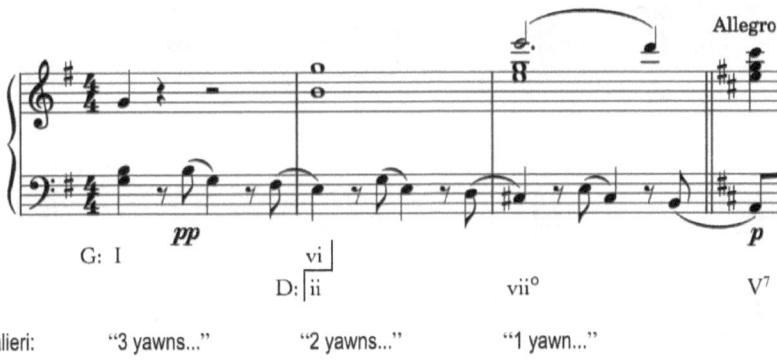

FIGURE 11. Salieri and Mozart discussing the failure of *The Marriage of Figaro*, in time to an excerpt of "Ah Tutti Contenti" (mm. 25–28).

stranger appears at his door and accuses him of "neglecting his request" (to compose the *Requiem*). Later, the "Ein Mädchen Oder Weibchen" aria from *The Magic Flute* cuts out on a V chord when a semiconscious Mozart is carried out of the concert hall (disc 2, 0:27:40), and the "Pa pa pa pa" aria is likewise arrested on a V chord as Salieri transports the ailing Mozart away in his carriage (disc 2, 0:28:33). Mozart's life and power is draining away, and Salieri swoops in to clinch the kill. To mark the triumphant accomplishment of Mozart's death, and add it to the trophy case containing his (Salieri's) father's death, the "Lacrymosa" from Mozart's *Requiem* is used analogously to the "Amen" from Pergolesi's *Stabat Mater*. The *Requiem* "Lacrymosa" accompanies Mozart's death and funeral, and the final "-men" syllable of "Amen" is synchronized with the visual return from flashback to present day (disc 2, 0:49:06) (see Video 6 ✶), mirroring the way the "-men" of "Amen" in the *Stabat Mater* brought Salieri back to the present after the flashback of his father's funeral (Figure 12, and refer back to Video 3 ✶). Salieri views these two deaths as crucial steps for advancing his career, and the magnificent cadences at the end of these works mark two momentous cadential accomplishments for Salieri.

But in a delicious twist of irony, fate erases Salieri's scheming endeavors and dooms him to *eternal* "unfulfillable longing." The concluding work of the film (the second movement of Mozart's Piano Concerto No. 20), which accompanies Salieri's quest for absolution (*"I absolve you!"*), is tauntingly festooned with Mozart's mocking laugh on the closing V^6_4 chord (disc 2, 0:51:19) (see Video 7 ✶). The PAC then occurs as the screen fades to black, symbolic of Salieri's fade into nothingness, as he realizes that history will remember Mozart's music and not his own (*and* the I chord arrives after Salieri has visually faded away, once again denying him cadential closure). Despite Salieri's success in snuffing out Mozart (Figure 13), Mozart in the end

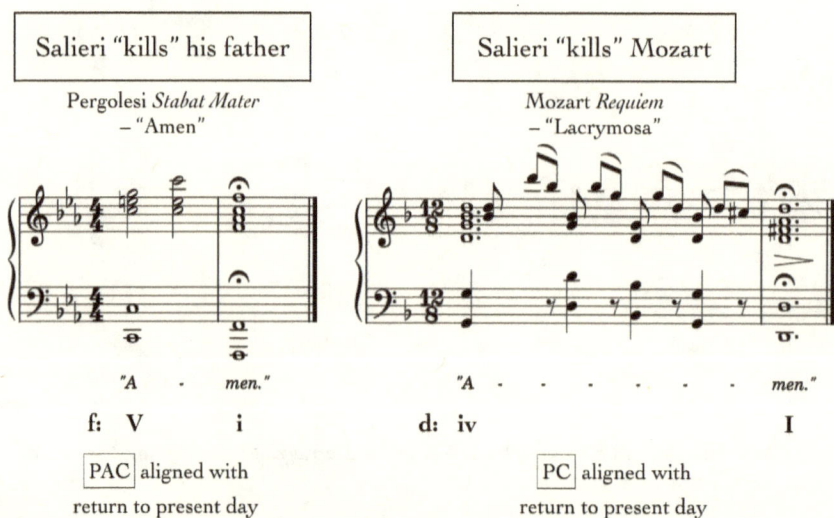

FIGURE 12. Salieri's murderous accomplishments, in which each final cadence is synchronized with a return from flashback to present day. Also, the massive minor-plagal cadence on the "Amen" of the "Lacrymosa" recalls the large-scale minor-plagal tonal motion celebrating the death of Salieri's father (see Figure 8), drawing a further connection between Salieri's two "murders."

triumphs over Salieri by spoiling his final chance of PAC resolution. Ending with that laugh of Mozart's is especially meaningful because earlier in the film Salieri had felt "that was *God* laughing at me, through that obscene giggle." He vowed at that moment: "Before I leave this earth, I will laugh at *You*" (1:08:34). Thus the fact that Mozart's (ergo God's) laugh is the final sound of the film—disrupting the final PAC—is symbolic of Salieri's cosmic defeat. This motif of cadential frustration as a symbol of "unfulfillable longing" is central to Salieri's narrative and illustrates the complex relationship between Salieri and Mozart in a subtle yet powerful way.

Because jump-cut tonality tends to be used most often in television shows, and are quick and simple to explain, I have grouped a few examples together in this short section.

30 ROCK, EPISODE 7.5 (2012)

Element featured: jump-cut tonality.

The highly caricatured diva Jenna Maroney (Jane Krakowski) begins singing the "Star-Spangled Banner" in D major at *nobody's* invitation (9:10), and a tonal jump

FIGURE 13. Early on in *Amadeus*, Salieri violently blows out a candle after promising God that he will someday vanquish Him/Mozart. Having repeatedly referred to Mozart as the Son of God, the three candles in this scene evoke the Holy Trinity, where Mozart—the Son—is the middle candle.

cut with the caption "Five Minutes Later" shows Jenna finishing the song in E major (see Video 8 ✱). The fact that she has modulated to a different key by the end emphasizes the inappropriate length of her performance (and the impatience of her audience), as incessant and unwelcome singing is one of Jenna's main character traits.

THE OFFICE, EPISODE 5.4 (2008)

Element featured: jump-cut tonality.

Corporate supervisor Jan Levinson (Melora Hardin) horrifies her employees by beginning to sing Dusty Springfield's "Son of a Preacher Man" at her office baby shower (11:40), which no one wanted to attend in the first place. Jan begins in D♭ major, and a tonal jump cut takes her to the key of F major, to illustrate how uncomfortably long she has been singing. A phone call between Jim and Pam (John Krasinski and Jenna Fischer) is spliced in between the tonal jump cut, to allow the audience to forget about Jan's song and thus be surprised and amused to find that she is *still* singing.

AT HOME WITH AMY SEDARIS, EPISODE 1.2 (2017)

Element featured: jump-cut tonality.

Absurdist Amy Sedaris, in this parody homemaker show, yells out "I'm in looooooooove!" beginning the long-sustained word "love" on a D, right before the commercial break (10:32). A tonal jump cut modulates her yell down to a B, humorously implying that she has been yelling during the entire commercial break and has followed the natural downward arc of a vocal shout.

FRIENDS, EPISODE 2.8 (1995)

Element featured: jump-cut tonality.

Ross Geller (David Schwimmer) commands his friend Phoebe Buffay (Lisa Kudrow) to sing a song, to ease the tension when both of his inamoratas (Jennifer Aniston and Lauren Tom) sit beside him in the coffee shop. The socially filterless and fearless Phoebe begins improvising a song about the love triangle, "Two of Them Kissed Last Night," in D major (4:14), making everyone uncomfortable because of her barely disguised references to the people present. A tonal jump cut takes Phoebe to an E-major chord, and she ends in A major. Phoebe's song began with a slow introduction (naming the three love-triangle characters), and the jump cut vaults her into an intense closing chorus (demanding that the lover man "must decide!"). These formal and tonal disjunctions make the song seem like an epic ballad that has been going on for the past twenty awkward minutes, augmented by the pained, uneasy looks on people's faces.

CLOSING THOUGHTS

Let me reiterate that the elements we have explored in this chapter are by no means an exhaustive list, and my hope is that readers will be inspired to find new ways in which keys create meaning in film. There are a host of other triadic transformations, for example, that might be used to create key relationships in soundtracks. For instance, in the next chapter, I briefly mention one casual example of a hexatonic pole relationship (in *Fantastic Mr. Fox*), but there are likely *many* other films that use this charismatic transformation to craft consequential key relationships. The "Tarnhelm" progression (minor-mode *Leittonwechsel*-Parallel, or **LP**, compound transformation, which transforms C minor into A♭ minor) is another harmonic relationship used ubiquitously in film music on the chord-to-chord level, which is likely to form the basis of key-to-key relationships as well.

Why might these types of elements be found in film soundtracks, we might wonder? For one thing, it's not an uncommon procedure for composers to sit at a piano/keyboard and block out chords as they map out themes and keys, and meaningful harmonic relationships can easily arise from the way block chords fit in the hand (especially transformational harmonies, which arise from parsimonious voice leading, or moving voices as little as possible). For another thing, Hollywood film music began as an extension of the nineteenth-century classical tradition, with early European émigrés reverently emulating Wagnerian techniques in their film scores. Although film music has evolved a great deal since the classic Hollywood style of lush European romanticism, Wagner still continues to inform film music, and many present-day film composers and music staff may have been exposed to his music to some degree in college music courses. So it's not hard to imagine how Wagnerian elements such as associative and directional tonality might find their way into film soundtracks. And the soundtrack as a cohesive entity (including original scoring, preexisting music, sound effects and dialogue), or *mise-en-bande*, has become increasingly more sophisticated over the past decades, which is why we can find elements of tonal design extending past original scoring and into the selection, arrangement, and manipulation of preexisting music as well as sound design (explored further in chapter 5).

Not every film will necessarily have an abundance of meaningful tonal elements to interpret, so don't force it. Once you start looking at the keys in a film, you will be inundated with potential key relationships (after all, labels exist for most harmonic relationships); the trick is to exercise analytical judgment (as discussed in chapter 1) in deciding which key relationships are salient with regard to the narrative, to create a compelling interpretation. Why do we have a tendency to notice traditional components like functional or relative relationships, when a film soundtrack is not *bound* by traditional tonal requirements? Because these types of common components form the foundation of our sonic world (in the West)—and hold meaning for us—so we as listeners are predisposed to notice them, just as filmmakers are predisposed to generate them. But, in interpreting film tonality, we are not *limited* to existing conventions, and citizen-analysts should always be on the lookout for novel tonal techniques suggested by the particular context of a film.

Of course, most listeners don't have perfect pitch (I certainly don't)—and even those who *do* cannot necessarily keep track of keys across an entire film; but like other forms of musical analysis, repeated hearings and analytical listening reveal new insights and layers of understanding. (After all, no one completes a metric analysis or sonata form interpretation of a Brahms symphony upon a first hearing or without taking copious notes.) Most of us grow up watching films as entertainment, boredom killers, or space fillers, with the idea that once we've seen a film (and know how it "turns out"), it is marked as COMPLETE on our checklist and not worth watching again. But like our favorite music, films should be experienced again and again for maximum engagement. The first viewing allows you to immerse yourself utterly in the narrative so that it sweeps you away; the subsequent viewings allow you to begin understanding and appreciating the craft and technique of the film.

Having now surveyed an assortment of films featuring most of the tonal elements discussed in the previous chapter, with one film loosely connected to the next, let's move on to chapter 3 in which *all* of the films can be thematically tied together by a particular tonal element.

3
Filmic Characters Rising Up and Settling Down

This chapter contains seven very different films, spanning half a century and ranging from stop-animation to historical drama; what connects these seven films is the presence of directional tonality in their soundtracks, which we can use as an interpretive lens for understanding the nuanced plights of their protagonists.

THE GRADUATE (1967)

Elements featured: associative tonality, transposition, directional tonality, functional tonality.

RELEVANT PLOT SYNOPSIS

Benjamin Braddock (Dustin Hoffman) is the affluent son of a Pasadena, California, high-society family, returning home for the summer after having just graduated from college. He feels listless, lonely, and directionless as he passively follows the life plan laid out for him by his parents. He finds himself drawn into an affair with a family friend, Mrs. Robinson (Anne Bancroft), even while his parents pressure him to date her daughter Elaine (Katharine Ross). He grudgingly takes Elaine on a date and does his best to repulse her, but his genuine interest is awakened once she calls him out on his boorish behavior. Benjamin and Elaine begin to bond as peers with similar problems and views (as he tires of his affair with her mother), and Mrs. Robinson is fiercely opposed to their relationship. She reveals that her accidental pregnancy (with Elaine) was what forced her into her unhappy marriage, and she threatens to expose their secret if Benjamin continues dating her daughter. Benjamin tells Elaine the truth about his affair, and she runs in horror to confront her mother, who retorts that Benjamin forced himself on her. Elaine cuts Benjamin off and returns to college but eventually accepts his side of the story after he follows her to Berkeley and doggedly pursues her. He asks Elaine to marry him, but she is already involved with a classmate, Carl Smith (Brian Avery). Elaine's parents pressure her into dropping out of college and marrying Carl, and Benjamin moves heaven and earth to crash the wedding. He arrives at the church mid-ceremony screaming "Elaine" to the bride's great relief. She runs toward him and Mrs. Robinson grabs her, yelling "It's too late!" but Elaine breaks free, yelling "Not for me!" Benjamin and Elaine sprint out the door and onto a bus, where their elation soon gives way to uncertainty, as they ride off into the unknown.

The Graduate (1967) is an iconic example of the surge of "compilation scores" in the late 1960s, which harnessed the logistical ease and financial profitability of constructing film soundtracks from preexisting pop songs.[1] While there are a few original jazz numbers (composed by Dave Grusin) used as background music in some scenes, preexisting Simon & Garfunkel songs form the backbone of *The Graduate*'s soundtrack. Director Mike Nichols selected these songs himself, to convey Benjamin Braddock's struggle to find himself and transition from boyhood to manhood.[2] Having just graduated from college, Benjamin is paralyzed with ennui to realize that he is destined to traverse the same suffocating life path as his parents. Benjamin begins the film by moving *toward* his parents and their life (flying back home and passively riding the moving walkway in LAX airport) to Simon & Garfunkel's "The Sound of Silence" in the key of E♭ minor (0:00:53). "The Sound of Silence" also accompanies the montage sequence in which Benjamin has his infamous affair with Mrs. Robinson—an affair that threatens to pull Benjamin further down the conveyor belt of bourgeois life (as he buries his dissatisfaction in life by getting a jump on the bored bed-hopping of the stultified suburban marriages in his world) (0:38:15). The third and final iteration of "The Sound of Silence" occurs at the very end of the film, when Ben rescues Elaine from her wedding and they ride away on a bus (1:44:48).

On the surface, this famous ending doesn't give us full assurance that Benjamin has actually succeeded in diverting the course of his fate away from the Pasadena life he wished to escape—after all, he *does* end up choosing the girl next door (as his parents had forcefully urged him), and his facial expression during this bus ride *does* bespeak a possible recurrence of ennui, being quite similar to his expression on the airplane at the beginning of the film (✱). But looking at the tonal design of this soundtrack gives us an intriguing clue: "The Sound of Silence" occurs three times throughout the film, and the final iteration is transposed *up* a half-step higher than the original key. Benjamin began the film by moving *toward* his parents and the Pasadena life (on an airplane) in the key of E♭ minor, and he ends the film by moving *away* from them (on a bus) in the key of E minor. This

---

1. See Hubbert (2014) for a concise history of the compilation scoring practice.
2. See Kashner (2008) for an account of Nichols's vision for this film.

surprising and significant E-minor transposition occurs after Benjamin has finally shed the mantle of boyhood and chosen his own path.

The tonal starting and ending points of this film thus form a meaningful arc of directional tonality, with the key rising upward to reflect the boy-becomes-man narrative. While Benjamin does end the movie much as he started, passively riding a moving conveyance, he made the *choice* to embark on this final conveyance. It may be the same song ("The Sound of Silence") and the same girl (from his family's caste)—but *the new key*, transposed upward (E♭→E), suggests that he has leveled up by making a stand, breaking free of his parents' control, and starting his own life. Benjamin's personal journey is one of awakening and uprising, and the upward transposition reflects the overcoming of his lethargy. We *see* this transformation visually—from Benjamin's comatose beginnings (sitting on an airplane, standing on a moving walkway, floating motionless in a pool) to his later animation (speeding and sprinting to chase after Elaine). And we *hear* this transformation tonally, rising directionally from E♭ to E minor, invoking the energy-gain of a large-scale pump-up modulation.

Elaine's character trajectory and its effect on Benjamin's, as well as an insight into their future prospects (revealed by an important sound effect), is discussed in chapter 5.

PERSUASION (1995)

Elements featured: associative tonality, transposition, directional tonality.

RELEVANT PLOT SYNOPSIS

Set in nineteenth-century England, Anne Elliot (Amanda Root) is the eldest daughter of an old aristocratic family, trapped in a torpid life that offers her no chance of progress or escape. Eight years ago, her family coerced her into refusing the proposal of her beloved Captain Wentworth (Ciaran Hinds), because he lacked fortune and status. Now twenty-seven years old, Anne is considered well past marriageable age, and the only respectable option remaining for her is to silently accept her permanent status as a spinster. Through the years of family oppression, Anne has grown downcast and meek, and she becomes a hollow shell of her former self.

Fate brings Captain Wentworth unexpectedly back into Anne's life. Having forged a successful career in the navy, he is now a wealthy and respected man, looking to marry and settle down. He still has strong feelings for Anne, but she is too despondent to realize it. During a group excursion to Lyme, Anne has the opportunity to reestablish the bond with him, but she is too weak-willed to respond. Captain Wentworth (still acutely raw over her past rejection) misinterprets her despondency for indifference and transfers his attentions to Louisa Musgrove (Emma Roberts), one of Anne's vain, silly cousins. Although Anne is heartbroken over this, the proximity to Captain Wentworth slowly brings her back to life. Captain Wentworth's foolish feelings for Louisa soon fizzle out, but meanwhile Anne is being pursued by another man (the heir to her family's fortune). Anne is blossoming from spending more time with people she truly cares about and less time with her abhorrent family members. Her confidence increases, and she grows stronger and more decisive. She finally asserts herself and shows Captain Wentworth that she still has feelings for him. Anne rebuffs intense family pressure to marry the rich heir and joyfully accepts Captain Wentworth's second marriage proposal instead. The ecstatic couple set sail on his ship, sailing off into the sunset with the whole world before them.

As in *The Graduate*, directional tonality adds an intriguing dimension to the interpretation of the narrative ending in this 1995 film adaptation of Jane Austen's *Persuasion*. Associative tonality also plays an important role, with the key of B major strongly linked to Anne Elliot's fate. Two harmonic sequences mark pivotal moments at the middle and end of Anne's narrative, and the key of B major within these sequences either thrives or is thwarted, depending on its relationship to the keys around it—symbolizing Anne's possible outcomes.

The first sequence occurs in the middle of the film, when Anne is afforded a miraculous second opportunity to win Captain Wentworth, who is still single and pining for her (and is now fabulously rich and distinguished). But once again, Anne proves too meek and dutiful to exert her own will over the forceful will of her disapproving family. This frustrating event is accompanied by a musical sequence (see Figure 15, further below) composed by film composer Jeremy Sams and inserted—as a very plausible middle B section—into Chopin's Prelude Op. 28, No. 3. The sequence (0:49:24) begins in E♭ major, modulates to G major, then to the somewhat unexpected key of B major, before returning to E♭ major.[3] The brief B-major opportunity presents itself, but irises shut when Anne fails to seize it. This sequence analogizes Anne's thwarted trajectory—her inability to break free from her constrictive life.

The second sequence occurs at the end of the film (1:44:46), when Anne and Captain Wentworth are finally reunited. This sequence (also composed by Jeremy Sams) *also* leads to B major, but this time—beginning from E major then G♯ major—the key of B major *ends* the sequence, and the cue, and the film. This time, the sequence outlines a major triad, rather than an augmented one, and the cycle does not complete itself, so the music does not return to its starting point (of E major). Instead, the film ends in B major as Anne and Captain Wentworth begin their exciting new life together. Thus, in this moment, B major is both a musical ending and a personal beginning. The achievement of B major symbolizes Anne breaking her cycle (of submissive obedience), overcoming the societal inertia keeping her in her place (socially sanctioned spinsterhood), and moving forward

---

3. Beginning with the keys E♭ major and G major, we might expect B♭ major to follow, since it would lead smoothly (via V-I motion) back to E♭ major. But instead of outlining a conventional major tonic triad, this sequence outlines a slightly less orthodox augmented triad and returns to E♭ major via a symmetrical major-third cycle.

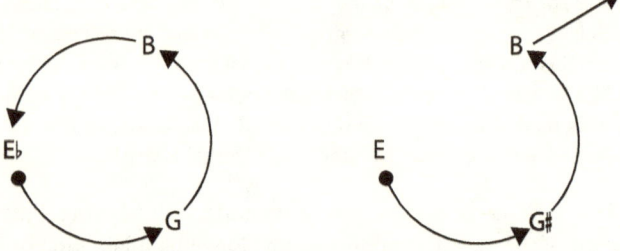

FIGURE 14. The two musical sequences in *Persuasion*.
Left: First musical sequence (middle of the film), in which Anne returns to the same place she began (E♭ major).
Right: Second musical sequence (end of the film), in which Anne breaks her cycle and moves forward on a new path (B major).

with the life of her own choosing. Figure 14 compares the trajectories of these two sequences side-by-side.

The key areas traversed in both sequences map onto physical locations Anne traverses during the scenes, in a way that reflects her outcome in each of these pivotal narrative moments. During the first sequence (Figure 15), Anne and Captain Wentworth (along with the rest of their party) walk along The Cobb pier, with the keys of E♭ major, G major, and B major corresponding (respectively) to the beginning, middle, and end of the pier. B major marks the moment they reach the end of the pier and look out onto the open expanse of the sea; but then they turn back and return to land (and E♭ major). Symbolically, B major is a briefly visited possibility that does not come to fruition, returning instead to E♭ major. In the second sequence (Figure 16), Anne sits below-deck on Captain Wentworth's ship (her new home) in E major, she moves up to the main deck in G♯ major, and she achieves B major when she climbs up to the quarterdeck to take her place beside her new husband—surveying the limitless sea before them.

The fate of B major within these two sequences depends on the tonal starting point. Proceeding from E♭ major (in the first sequence), B major fails to take hold; its presence is rather unexpected in the context of E♭ major and G major, outlining the uncomfortably uncanny augmented triad, and it resolves abruptly back to E♭ major.[4] Analogously, the unexpectedness of B major in the context of the E♭-major sequence is suggestive of the very surprising second chance Anne has with Captain Wentworth. In this cultural time and place a man like Captain Wentworth would have long ago found some other young lady to marry, and a woman like Anne would have been relegated to the permanent role of maiden aunt to her

---

4. For a discussion of the uncanny nature of the augmented triad, see Cohn (2012).

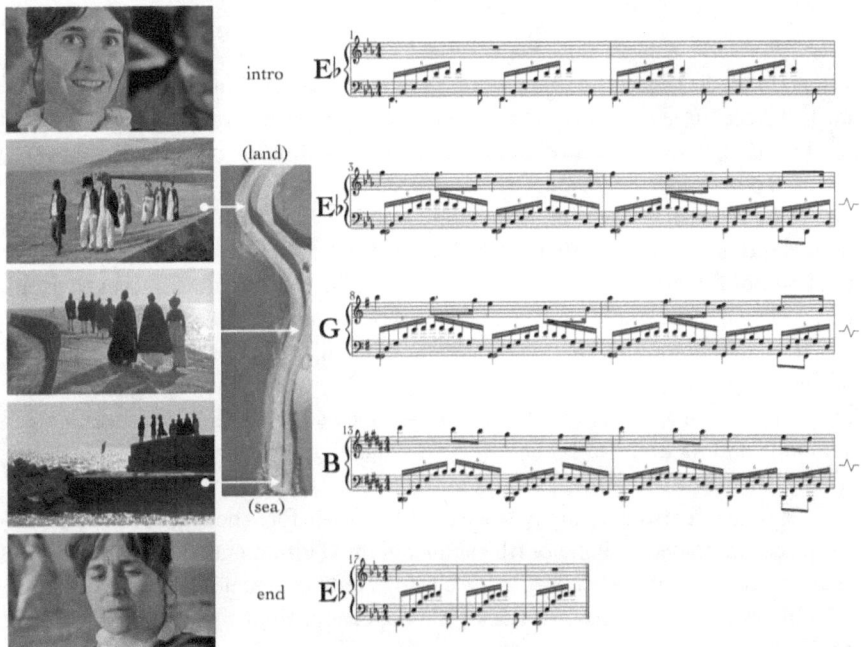

FIGURE 15. Physical locations along The Cobb at Lyme Regis Harbor (photo from Google Maps), corresponding to key areas in the first musical sequence, inserted into Chopin's Prelude Op. 28, No. 3 by Jeremy Sams. (Break-lines indicate omitted measures.)

FIGURE 16. Physical locations on the ship, climbing successively higher with each key area of the musical sequence in the "Fulfillment" theme, composed by Jeremy Sams. (Break-lines indicate omitted measures.)

sisters' spoiled children. In other words, E♭ major was not "supposed to" lead to B major under these societal circumstances. And since Anne has not yet mustered up the courage to take control of her life at this point in the narrative, B major is subsumed back into E♭ major.

But by the end of the film, Anne and Captain Wentworth have overcome the obstacles that lay between them, and B major is securely achieved from the starting point of E major. The moment she and Captain Wentworth lock eyes to form their covenant, the music of a passing circus procession engulfs them in E major, and from this point forward, the remainder of the film (all breathless anticipation of their impending marriage) is underscored by E major.[5] Beginning a half-step higher from E major (rather than E♭ major) allows the second sequence to lead smoothly and naturally to B major (via G♯ major), outlining a harmonious major triad.[6] *Now* Anne is ready to proceed into the open waters of her new life, and she and Captain Wentworth survey the expansive sea before them and sail into the sunset accompanied by B major (the ending of the "Fulfillment" theme plus the "O Mio Tesoro" cue that plays during the closing credits). B major represents Anne's blissful new life, and E major is the key that opens the door to her B-major happy ending/beginning.

The keys of E♭ major and G major (in the first sequence) and E major and G♯ major (in the second sequence) are not used associatively in this film—that is, they don't carry any specific meaning in the narrative—but are used logistically as to lead Anne toward or away from the associatively significant key of B major. B major is subtly intimated twice before, earlier in the film, hinting at its important function later in the film. In the first instance, Anne wistfully tinkers on the piano in B major while thinking about Captain Wentworth (0:19:29). (Immediately preceding this, Louisa plucks the pitches B-G♯-F♯ on her harp, which provides a $\hat{1}$-$\hat{6}$-$\hat{5}$ lead-in to Anne's B-major tinkering.) The second instance occurs when Anne and Captain Wentworth listen to a love aria ("O Mio Tesoro") in B major at a salon recital (1:23:42) and reignite their passion for one another. Both B-major hints point Anne in the direction of Captain Wentworth, even though it takes her a while longer to finally clinch B major for herself.

When analyzing the *Persuasion* soundtrack, it is intriguing to notice that *all* the preexisting music has been transposed to new keys. This film utilizes several Chopin and Bach piano pieces, *none* of which appear in their original keys (Table 4). This conspicuous characteristic invites interpretation, since the transposition of

---

5. A marching band plays as Captain Wentworth and Anne make their covenant (1:39:47), a solo piano plays as they bask in their love (1:41:59), and an orchestra plays as they sail off on their ship (1:44:46). All three are separate themes, unnamed in the credits or soundtrack listing.

6. As V of E major, B major is a "natural" choice (forgive the pun) that requires no accidentals and is a conventional and expected tonal destination.

TABLE 4 Preexisting works in *Persuasion*, all transposed to new keys for the film

Cue	Original Key	Film Key	Transposition	Timestamp
Chopin Prelude Op. 28, No. 21	B♭ major	A major	down m2	0:00:15
Bach French Suite No. 1—III. Sarabande	D minor	C♯ minor	down m2	0:14:04
Chopin Nocturne Op. 9, No. 3	B major	B♭ major	down m2	0:41:52
Chopin Prelude Op. 28, No. 3	G major	G♭ major	down m2	0:49:24
				0:51:35
				0:54:28
Bach French Suite No. 3—III. Sarabande	B minor	C♯ minor	**up** M2	1:16:16

preexisting works requires an intentional effort (i.e., the digital manipulation of existing recordings or requiring a pianist relearn these canonical works in new keys). These are all unaccompanied solo piano works, a genre closely identified with Anne because her unaccompanied solo (single) status is the central theme of the narrative, and because she is shown playing the piano all throughout the film.[7] As seen in Table 4, the first four works—spanning the first three-quarters of the film—are transposed *down* a half-step from their original keys, and the half-step lowering of these musical selections is illustrative of Anne's depressive state.[8] Anne's low spirits are a frequent topic of discussion among her friends and family, she meekly allows herself to be downtrodden by everyone around her, and her lowly state is even reflected in her physical appearance (with her drooping posture, limp hair, and pallid cheeks). But as Table 4 shows, the final preexisting work in the film is transposed *up* a whole-step from its original key—and this corresponds to the time when Anne finally regains her spirit and begins making her own choices.[9]

It is significant that this final upward transposition is a whole-step rather than a half-step: an upward half-step would simply cancel out the initial half-step depression, doing no more than undoing the wrongs inflicted by Anne's family, narratively speaking. Instead, the whole-step uprising implies that her reward goes beyond that

---

7. Also, well-bred young ladies in Victorian England were the main purveyors of solo piano music, and an unmarried country gentlewoman without nieces and nephews, such as Anne, would have spent a great deal of her waking hours at the pianoforte.

8. Another affirmation for this interpretation of the half-step pitch depression can be gleaned from the starting point of the two musical sequences: the thwarted-trajectory sequence begins a half-step lower (E♭ major) than the triumphant-trajectory sequence (E major), and it is this lowered half-step that makes the key of B major tonally unachievable and unsustainable at that point in Anne's narrative.

9. For instance, she begins spending more time in the company of the Wentworths, and she eschews the exalted visitation of Viscountess Dalrymple to instead visit her low-ranked friend, Mrs. Smith—all despite the vehement disapproval of her family. Anne starts holding her head up higher, taking more care with her hair and dress, speaking in a more confident tone, and the color returns to her cheeks.

simple correction. In many ways, Anne is exceeding the happiness she would have experienced in marrying Captain Wentworth when they were young: he is now independently wealthy and socially established (which Jane Austen does consistently posit as substantial prerequisites for a happy life), they begin their life together *after* Wentworth fought through the Napoleonic Wars (which would have been a perilous time for a new marriage), and they are able to quietly sever ties from her toxic relatives (which wouldn't have been possible earlier in Anne's life).[10] This whole-step happy ending tells the tale of middle-aged love triumphant, in which the lovers achieve greater happiness for having earned it through patience and endurance.

The whole-step uprising also manifests itself in the tonal envelope of this film. Comparing the tonal starting point (A major) and ending point (B major) of *Persuasion*, we see another example of the whole-step uplift, illustrating Anne's narrative trajectory by means of directional tonality. Like Benjamin in *The Graduate*, Anne needed to rouse herself from stupor in order to overcome the repression of her oppressive family and forge her own life path—so the tonal direction for both protagonists is *upward*.

MOONLIGHT (2016)

Elements featured: associative tonality, transposition, directional tonality

RELEVANT PLOT SYNOPSIS

This tripartite film follows the development of a young gay Black man from childhood ("Little") to adolescence ("Chiron") to manhood ("Black").

I. LITTLE

The sensitive, withdrawn child of a drug-addicted single mother (Naomie Harris), Little (Alex Hibbert) accidentally stumbles onto the parental figures he craved, Juan (Mahershala Ali) and his girlfriend Teresa (Janelle Monáe). Little attaches himself to these two nurturing role models, who (in response to his inquiry) are the first people in his life to affirm that it's okay to be gay. Juan teaches Little about life and manhood, but his mother (one of Juan's crack customers) is jealous of their relationship. Juan berates her for smoking crack and neglecting her son, but she defensively blames Juan for selling her the crack in the first place.

II. CHIRON

In his teen years and using his given name, Chiron (Ashton Sanders) struggles with a school bully (Patrick Decile) who torments him for being different. Juan has died and

---

10. The film begins just after Napoleon was exiled to Elba and ends some months later, meaning that Napoleon's escape and subsequent mischief will occur soon after Wentworth and Anne marry. However, the new couple stands a far better chance of weathering the ensuing Hundred Days War than the previous eleven-year battle, during which Anne would most certainly not have been allowed to accompany her husband.

FIGURE 17. Main theme of *Moonlight*.

Chiron continues relying on Teresa for support, as his mother's addiction has caused her life to unravel further. Chiron is enamored with his childhood friend Kevin (Jharrel Jerome), and the two share a secret romantic encounter (Chiron's first). The next day at school, the bully convinces Kevin to publicly attack Chiron. Chiron avenges himself against the bully the following day, and he (Chiron) is arrested.

III. BLACK

As a young man, now going by the name of Black (Trevante Rhodes), he has established a comfortable and stable life for himself as a drug dealer. Black forgives his mother, who now lives in a drug rehab center, and makes a trip to visit Kevin (André Holland). The story ends with the renewal of friendship between Black and Kevin, and implies that they are beginning a romantic relationship.

Moonlight, like *The Graduate*, portrays the coming-of-age of a young man. This subtle and powerful film by director Barry Jenkins features a tonally expressive soundtrack in which keys respond sensitively to the stages of this character's life. At the heart of the soundtrack is a tender theme composed by Nicholas Britell to serve as the protagonist's leitmotif (Figure 17).

The first time we hear this theme (0:06:52), it is in the key of D major and billed (in the soundtrack listing) as "Little's Theme." Once Little sheds his childhood nickname and transitions to being called Chiron, the theme (now called "Chiron's Theme") is transposed downward to B major (0:36:55). Finally, as a grown man nicknamed Black, the theme (titled "Black's Theme") is transposed further downward to A major (1:39:22). These three chapters of the character's life are visually delineated with title cards reading "i. Little," "ii. Chiron," and "iii. Black," and aurally delineated by the keys D major, B major, and A major (respectively). While the character keys of D major and B major appear within a minute of the corresponding title cards, Black's key of A major appears thirty minutes after the "iii. Black" title card. This is because the three pivotal keys represent the three developmental phases of the protagonist's life, and Black does not fully attain his sense of self and manhood until later in the "iii. Black" chapter.

After the official introduction of this chapter, we hear another iteration of the B-major theme (in this instance branded as "Atlanta Ain't But So Big") (1:10:56) as Black seems to be going through the motions of adult life, dealing drugs and drifting in a direction he didn't seem to want for himself as a youth (having loathed what drugs did to his mother). It's not until his close childhood friend Kevin calls him out and brings him back to himself that Black finally settles into adulthood, with the "Black's Theme" in A major (1:39:22). This A-major moment arrives when Black looks out at the ocean, feeling a powerful sense of homecoming—in more ways than one. The ocean was where he experienced the two most important moments of his childhood and adolescence: the day that Juan taught Little how to swim and imparted formative life lessons, and the day that teenaged Kevin gave Chiron his first sexual experience and taught him it was okay to be himself. Thus, when Black gazes at the ocean for the first time since these two momentous events, the newly deepened A-major theme signals that he has officially matured into his adult self. The continual downward transposition of Britell's plaintive theme, sung first by a violin then a cello, beautifully follows the deepening of this character's voice as he matures from boy to man.

In addition to the downward transposition of Britell's original theme, certain preexisting works are also transposed downward for this film (the transposition technique used in this film is discussed in chapter 6). Erykah Badu's B♭-minor "Tyrone" is transposed down to F minor for its use in this film (1:08:36), and Jidenna's G-minor "Classic Man" is transposed down to E minor (1:22:06 and 1:37:00). Both of these songs occur during the final "iii. Black" chapter (the Badu immediately after the chapter title card is shown), and they are transposed downward by slowing down and warping the sound. The slowed, lowered voices in these transpositions aptly evoke the tone of Black's adult voice (which is why Badu's female voice must be downward-transposed by a larger interval than Jidenna's male voice). Boris Gardiner's "Every Nigger Is a Star" is an especially interesting transposition due to its position as the very first cue of the film—I address it at the end of this analysis.

C major acts as a key of redemption in this film, occurring only in two impactful moments during the last chapter of the film when Black reestablishes contact with his mother and Kevin, accepting their amends for the ways they had hurt him. Tomás Méndez's "Cucurrucucú Paloma" in C major plays prominently after Black makes peace with his mother (1:20:58), forgiving her for the neglect and abuse he suffered as a result of her drug habit. And Kevin plays Barbara Lewis's C-major "Hello Stranger"—their special song from long ago—on the jukebox as a reconciliation (1:34:52), after their rift of many years (teenaged Kevin had begrudgingly but brutally attacked Chiron at the behest of the school bully). This last C-major song is also preceded and followed by C-pitched sound effects: a C-pitched ping as Kevin drops the coin into the jukebox and a C-pitched car horn when Kevin and Black leave the diner together. In a soundtrack that traverses virtually every key, C major is conspicuously earmarked for these climactic, redemptive moments in Black's narrative.

D major, B major, and A major are the character-defining keys in *Moonlight*, representing the three phases of Little's/Chiron's/Black's life—just as the film's cover art represents his face as a composite of the three facets of childhood, adolescence, and adulthood. The use of sound effects in supporting these keys is mentioned in chapter 5.

The transposition of *Moonlight's* opening cue, Gardiner's "Every Nigger Is a Star" (0:00:05), sets the direction for the tonal envelope of this film. The original recording of the song is in D major, but the film transposes it down a half-step, so that the film begins in C♯ major instead. Had the song remained in its original key, the *Moonlight* soundtrack would have exhibited tonal closure, beginning and ending in D major (the final two cues of the closing credits being in D). As it is, this purposefully transposed starting point forms an overarching trajectory to the film soundtrack, with the directional tonality of the C♯→D bookends conveying the direction of the main character's life. On one level the key of the Britell main theme lowers as Little/Chiron/Black ages, to reflect his deepening voice; but at the larger scale his tonal trajectory is upward, as he evolves from the passivity of childhood to the assertiveness of adulthood.

Beneath its numerous masterfully woven threads of narrative motifs, *Moonlight* is a love story; I proffer the film's main musical theme in support of this argument. Britell's poignant theme (see Figure 17) features the "Heartstring Schema" (a I–X–I progression where X is a chromatic chord containing ♭$\hat{6}$), which has been used, from the nineteenth century through the present day, to portray a bittersweet sense of romantic yearning (Schneller and Motazedian 2017). Scoring practice for classic Hollywood love themes made almost formulaic use of the Heartstring Schema, so its central presence in the main theme of *Moonlight* subtly positions this film as a romance.[11] Like the previous romance we analyzed, *Persuasion*, a central thread of *Moonlight* is the protagonist's defiance of familial constraints and social conventions in pursuit of non-normative bliss. Both films feature a great deal of transposition to portray the protagonist's personal journey, and both films feature a directional uplift in the tonal envelope to reflect the attainment of a "happy ending." Anne Elliot rises up to defy her age-sanctioned spinsterhood and caste requirements for marriage, while Little/Chiron/Black rises up to defy the

---

11. So ubiquitous was the Heartstring Schema in early Hollywood love music that Frank Skinner, in his 1960 film scoring manual, *Underscore*, demonstrates two different Heartstring chords (♭VI and CT°7) in his illustration of the exemplary "theme for an intense dramatic love scene" (Schneller and Motazedian 2017).

gravitational forces of poverty, drug use, and cultural homophobia.[12] Quiet, calm, and old-souled in demeanor, both of these characters wait till a slightly later-than-usual life stage to stand up for what they want; but in the end, Anne chooses her desired man (and life), and Black does the same.

THE ROYAL TENENBAUMS (2001)

Elements featured: associative tonality, transposition, singular key, parallel relationship, directional tonality, tragic-to-triumphant arc, meta key.

RELEVANT PLOT SYNOPSIS

The Tenenbaums are a dysfunctional family of unhappy individuals. The three adult children were extremely gifted when they were young: Chas (Ben Stiller) was a finance genius, Margot (Gwyneth Paltrow) (who was adopted) was an award-winning playwright, and Richie (Luke Wilson) was a professional tennis star. But Royal (Gene Hackman) and Etheline (Anjelica Huston) separated when the children were young, and Royal became estranged from the family.

Decades later, all of the Tenenbaum children have failed as adults and are heavily depressed. Chas is an angry widower, crippled by the death of his wife and his subsequent paranoia to keep his sons safe. Margot hasn't written anything for years and is bored in her marriage to Raleigh St. Clair (Bill Murray). Richie is aimlessly traveling the world, following the public meltdown of his tennis career.

Out of the blue, Etheline's longtime accountant and family friend, Henry Sherman (Danny Glover), asks her to marry him. Learning of Henry's proposal, Royal becomes jealous and territorial. Broke and recently evicted, Royal pretends to be dying of cancer to coerce his family's sympathy, and he moves back into the family home to try and win back their affections. Royal makes unsuccessful efforts to reconnect with each of his children and his two grandsons, and he attempts to come between Etheline and Henry. Henry is suspicious of Royal and eventually exposes to the family that Royal has been lying about having cancer. The family is disgusted to learn of Royal's deception, and he is shunned once again.

Troubled by Margot's increasingly secretive behavior, Raleigh and Richie hire a private investigator to spy on her. The PI reveals the shocking truth about Margot's secret life, which includes (among other things) compulsive infidelity. Her husband Raleigh is crushed, but her (adopted) brother Richie is even more distressed. He has been secretly in love with Margot his whole life, and he is shattered to learn of her betrayal. He attempts suicide, and Margot (realizing he did it because of her) is devastated. Richie's suicide attempt is a wakeup call for the Tenenbaum family. Margot admits to Richie that she loves him too. Etheline realizes she wants to marry Henry. And Royal realizes he needs to do right by his family for once in his life: he finally grants Etheline divorce papers (so that she can marry Henry), and he repairs his

---

12. Black may deal drugs, but he does not *do* drugs, which is what ruined his mother. He even makes a point of avoiding alcohol and tea, as he mentions in Kevin's diner.

relationships with Richie and Margot—though Chas is still too angry to forgive him at this point.

Before Henry and Etheline's wedding, Eli Cash (Owen Wilson), the Tenenbaums' childhood friend and neighbor, crashes his car into their house, while high on drugs. Royal rescues his grandsons from being crushed by the car. This triggers a cathartic outburst in Chas, and both he and Eli finally admit they need help. Chas is grateful to Royal for saving his sons and finally forgives him.

After this, all of the Tenenbaums are on the path to leading happier lives. Etheline marries Henry, Margot releases a new play, Raleigh publishes a new book, Eli checks himself into a drug rehabilitation facility, Richie begins teaching a junior tennis program, and Chas becomes a more relaxed father. After having repaired his relationship with his family (and their relationships with one another), Royal has a heart attack and dies peacefully. The family attends his funeral and are all finally at peace with themselves and one another.

The Royal Tenenbaums opens by drawing the audience into a book (✱), and the music and sound effects help transport the viewer-listener into the diegesis (see Video 9 ✱). The F-minor opening music ("111 Archer Avenue" theme, written by the film's composer, Mark Mothersbaugh) begins by shimmering on a dominant-ninth chord (C-E-G-B♭-D♭) during the "Touchstone Pictures" logo and fades away as the opening scene reveals the eponymous book on a library circulation desk. The librarian reaches out to stamp the book's card as a desk bell pings on the pitch C, emphasizing $\hat{5}$. A voice whispers "shhhhh," signaling that a story is about to be told, the librarian turns the book around to face the patron (us), and the V^9 finds resolution in the tonic F-minor chord of the music, just as the opening credits appear across the screen. This cue ends once again on a V^7 chord, as the book's "Prologue" page is opened to us. We officially enter the book's narrative in the next scene, when the austere Tenenbaum house is shown, and the preceding V^7 chord finds tacit resolution in the opening tonic chord of The Beatles' "Hey Jude" in F major. So the dominant seventh chord is used to introduce us to the book, and then to the narrative *within* the book. In fact, the cue that accompanies the Prologue (the second movement of Ravel's String Quartet) ends on a dominant chord as well, when the Prologue is over, leading us into the narrative proper: "Chapter One," which begins in present day, after the Prologue's summary of the past.

The overall tonal framework of this film follows a tragic-to-triumphant arc, beginning in F minor and ending in F major, which mirrors the Tenenbaum family's teleological arc. F minor represents the state of disunity and dysfunction in which they begin, and F major exemplifies their reunion and redemption (Table 5). Two montage sequences at the start of the film illustrate this dichotomy: Enescu's F-minor Cello Sonata accompanies a montage of the adult characters' maladjusted lives, while The Beatles' F-major "Hey Jude" recounts the promising potential they once had as children. All of the F-minor music occurs within the first

TABLE 5 F-minor and F-major moments in *The Royal Tenenbaums*

F-minor Cue	Type	Plot Event	Timestamp
Mothersbaugh "111 Archer Avenue"	original	book is introduced	0:00:03
Enescu Cello Sonata—I. Allegro molto moderato	preexisting	unhappy, isolated adults are introduced	0:08:12
Mothersbaugh "Look at That Old Grizzly Bear"	original	Royal feels jealous (of Henry) and rejected (by Etheline)	0:30:03

F-major Cue	Type	Plot Event	Timestamp
The Beatles "Hey Jude"	preexisting	montage of the children's prodigious talents	0:00:44
Vince Guaraldi Trio "Christmas Time Is Here"	preexisting	Margot moves back home	0:20:12
Nico "These Days"	preexisting	Margot and Richie reunite	0:25:13
Vince Guaraldi Trio "Christmas Time Is Here"	preexisting	Royal and Margot reconcile	1:27:33
Mothersbaugh "Rachel Evans Tenenbaum"	original	Royal wins his family's acceptance by making amends	1:29:09
Mothersbaugh unnamed theme #5*	original	closing credits	1:46:03

* There are five unnamed themes in this film, clearly composed by Mothersbaugh, which are not included in the credits, commercial soundtrack album, or acknowledged elsewhere.

thirty minutes of the film, and F major reigns in the last thirty minutes of the film. The closing credits feature an unnamed theme in F major (clearly written by Mark Mothersbaugh), which combines elements from several other Mothersbaugh themes that were all (but one) originally in other keys. This F-major theme serves to enclose the entire film (from opening credits to closing credits) in the tragic-to-triumphant envelope. So the choice to end the closing credits with this newly composed F-major theme, rather than simply reuse one of the existing themes (as is commonly done in closing credits), implies a deliberate decision to end the film in F major.

Within the F minor-to-major frame the other main key areas of the film are A minor/major and C major. A minor is used to illustrate particularly dark moments in their lives, while A major accompanies two rare moments of pure joy (Table 6). C major underscores moments of bonding and intimacy, in which these estranged family members lower their guard and connect with one another (Table 7). F, A, and C outline an F-major triad, alluding to an F-major meta key in *The Royal Tenenbaums*.

Among the smattering of other keys found in this film, one of them stands out among the rest: E♭ major is the key that underscores the astonishing revelation of

TABLE 6 A-minor and A-major moments in *The Royal Tenenbaums*

A-minor Cue	Type	Plot Event	Timestamp
Ravel String Quartet—II. *Assez vif*	preexisting	un-narrated visual introduction to adult characters	0:06:46
Mothersbaugh unnamed theme #1	original	Etheline discusses Royal's (fake) cancer	0:23:48
Mothersbaugh unnamed theme #2	original	Royal discusses his (fake) cancer	0:27:20
Elliott Smith "Needle in the Hay"	preexisting	Raleigh's depression and Richie's suicide attempt after the revelation of Margot's shocking secret life	1:09:07
The Clash "Rock the Casbah"	preexisting	Eli's drug addiction	1:25:14
A-major Cue	Type	Plot Event	Timestamp
John Lennon "Look at Me"	preexisting	Chas settles the boys into bed, and they all feel secure and happy for the first time in a long time	0:17:39
Paul Simon "Me and Julio Down by the Schoolyard"	preexisting	Royal takes the boys out for a day of pure reckless fun	0:51:30

TABLE 7 C-major moments in *The Royal Tenenbaums*

C-major Cue	Type	Plot Event	Timestamp
Bob Dylan "Wigwam"	preexisting	Etheline and Henry's first kiss, and Royal's first contact with the boys	0:32:34
Mothersbaugh "Mothersbaugh's Canon"	original	Royal bonds with Richie, and Margot bonds with the boys (at the cemetery)	0:35:29
Emitt Rhodes "Lullabye"	preexisting	Royal's first night back in the house with the family	0:46:08
Mothersbaugh "Scrapping and Yelling"	original	Royal and Etheline reconnect (walking in the park)	0:54:53
The Rolling Stones "She Smiled Sweetly"	preexisting	Richie and Margot bond (in his tent)	1:17:57
The Rolling Stones "Ruby Tuesday"	preexisting	Margot tells Richie their love must remain secret	1:20:46
The Velvet Underground "Stephanie Says"	preexisting	Mordecai returns to Richie (after decades away)	1:24:03
Mothersbaugh "Sparkplug Minuet"	original	Eli and Chas both admit they need help, Royal and Henry make peace, Royal buys a new dog for the boys, Royal bonds with Chas	1:34:40

Margot's secret life. Because this key is used only once in the film (The Ramones' "Judy Is a Punk" [1:07:45]), we cannot definitively designate this as an associative key (as discussed in chapter 1). But we can certainly speculate that a singular key was used in this context *because* it is such a singular event—and the crucial crisis that causes the Tenenbaum walls to come tumbling down. The volume of this E♭-major cue is significantly louder than anything else in the film, so it demands our attention *sonically* as well as tonally. There is only one other singular key in this film, and it too is directly related to Margot's shocking secret (Nick Drake's "Fly" in A♭ major is playing when Richie returns to Margot after his suicide attempt, which was brought on by the E♭-major discovery of her secret life [1:16:06]). So we have good reason to interpret this singular use of the key of E♭ as an element of the film's tonal design.

Within this film's tragic-to-triumphant F minor-to-major trajectory, there is another significant tonal element embedded. The final cue of *narrative* space (just before the closing credits) is Van Morrison's "Everyone" in G major, the lyrics of which perfectly capture the denouement of the Tenenbaum family's narrative arc:

> We shall walk again down along the lane,
> Down the avenue just like we used to do,
> Everyone, everyone, everyone, everyone!

From the morose F-minor beginning of the film to this jubilant G-major celebration, we can interpret an arc of directional tonality, as these characters finally break out of the catatonic lethargy of stunted pre-adulthood and move on with their lives. And if we dig even further, the previous two cues before the G-major "Everyone" rise continually upward in key (Table 8), making the rising tonal trajectory palpable on a more localized scale.

So while the F minor-to-major tonal envelope of the film represents the estranged-to-reunited trajectory of the family *as a whole*, the interior F minor to G major envelope shows the personal growth of each *individual* family member. This is directional tonality embedded within a larger tragic-to-triumphant arc.

HIDDEN FIGURES (2016)

Elements featured: transposition, directional tonality.

Hidden Figures tells the uplifting true story of the first Black female mathematicians working at NASA during the 1960s space race and their persistent endeavor to rise upward within the institution. Without having done a full tonal analysis of the film, I did immediately notice the presence of transposition and directional tonality in this soundtrack. A musical theme in D minor begins when Katherine Goble Johnson (Taraji Henson) is shown hard at work as the sole Black female

TABLE 8 Final three cues in *The Royal Tenenbaums*, leading into closing credits (just before the final F-major cue)

Cue	Key	Plot Event	Timestamp
Mothersbaugh "I Always Wanted To Be a Tenenbaum"	E major	Margot finally speaks honestly and shares her secrets	1:38:03
Nico "Fairest of the Seasons"	F♯ major	Each character's resolution (and future) is shown	1:39:22
Van Morrison "Everyone"	G major	Everyone leaves Royal's funeral feeling free (closing credits begin)	1:42:29

member of the Space Task Group (0:56:39).[13] Later, this theme is transposed up a whole-step to E minor when Dorothy Vaughan (Octavia Spencer) confidently leads the entire Black female computing team from their segregated quarters across the NASA campus to join the rest of the data center staff (1:27:51). This triumphant directional transposition sonically illustrates the uprising of the Black women of NASA (see Video 10 ✴).

These first five films have all featured *upward* directional tonality, representing the characters' successful personal trajectories. The last two films discussed in this chapter feature *downward* directional tonality, which illustrates a different way of achieving success.

EMMA (1996)

Elements featured: associative tonality, transposition, directional tonality, functional harmony.

RELEVANT PLOT SYNOPSIS

Set in the nineteenth-century English countryside, Emma (Gwyneth Paltrow) is a beautiful, wealthy young lady with too much leisure and not enough sense. Having nothing of substance to occupy her mind, Emma idles her time away with childish, frivolous pursuits. When she happens to predict the marriage of two friends, she

---

13. This theme—composed by Hans Zimmer, Pharrell Williams, and Benjamin Wallfisch—is not listed in the end credits or the official commercial soundtrack album, which features just the (vocal) songs written by Pharrell Williams alone. The unofficial soundtrack labels this theme "Call Your Wives" (D-minor version) and "Ladies' March" (E-minor version) even though the iterations are virtually identical (other than key). "Call Your Wives" even modulates from D minor to E minor halfway through, basically replicating "Ladies' March." As is common with underscoring track listings, these theme names are essentially chapter titles in the film, reflecting what is happening in the narrative during these scenes.

TABLE 9 Theme key transpositions in *Emma*

Theme	Keys
"Main" theme	A major
	G major
	F major
	G minor
	F♯ minor
	E minor
	D major
"Activity" theme	G major
	A major
"Mischief" theme	D minor
	G minor
"Tension" theme	D minor
	G minor
"Elton" theme	G major
	D minor

suddenly fancies herself a matchmaker, and it becomes her new favorite pastime. Her lack of life experience combined with her quixotic judgment means that she is *never* right in any of her hunches, and she makes comically misguided attempts to find love for her friends and neighbors.

For her own part, Emma imagines that she is impervious to love and matrimony. But when her protégé Harriet (Toni Collette) falls in love with her lifelong best friend, Mr. Knightley (Jeremy Northam), Emma suddenly realizes that she herself is in love with Mr. Knightley. In her usual jumbled fashion, Emma misreads Mr. Knightley's feelings and thinks that he loves Harriet. Feeling jealous and petulant, Emma ungraciously humiliates an elderly friend. Mr. Knightley, who is *also* feeling jealous and petulant (because he mistakenly thinks that Emma is in love with the foppish Frank Churchill [Ewan McGregor]), gravely berates Emma for her bad behavior. Emma is devastated by Mr. Knightley's disapproval and strives to better herself, in order to become the woman he hopes she can be. With his ideal as her guide, Emma matures into a (slightly) wiser and more sensible version of herself. As a result, Emma and Mr. Knightley finally acknowledge they are in love with one another and are happily united in marriage.

Like *Persuasion*, the 1996 film adaptation of Jane Austen's *Emma* features a great deal of key transposition, but in this case, involving original themes (rather than preexisting works), as in *Hidden Figures*. The *Emma* score consists of five different themes composed by Rachel Portman (✶), and every one of them appears in multiple keys (Table 9).

TABLE 10 Important key areas of "Main" theme in *Emma*

Key	Plot Event	Narrative Location	Timestamp
A major	Emma is introduced	beginning	0:00:15
D major	Emma and Knightley declare their love	climax	1:50:34
G major	Emma's story ends (with her wedding)	ending	1:57:00

Most of the themes occur in two keys, but the "Main" theme is transposed to seven different keys. While associative tonality is not a major factor in this film, we can glean meaning from one of the keys in which the "Main" theme appears. The most important iteration of the "Main" theme occurs in D major, at the teleological climax of the film, when Emma and Knightley declare their love to one another (1:50:34). This is the only time any of Portman's original themes occur in the key of D major. This in itself might not draw our attention, but there is one other instance of D major in the entire film, and it corroborates the romantic nature of this key. Earlier in the film, the chapel organ plays recessional music in D major while everyone exits the chapel (1:09:57)—the very chapel from which Emma and Knightley will be married at the end of the film. This D-major organ music thus acts as a subtle herald of their impending union. The chapel organ itself (as an instrument) is somewhat evocative of marriage, as wedding music is one of its main functions in film; thus, pairing it with D major in this instance strengthens the marital insinuations associated with this key in *Emma*.

The film opens with the "Main" theme in A major (0:00:15) and ends with the "Main" theme in G major (1:53:36). Rather than begin and end with the same theme in the same key (which would certainly be the easier logistical option), the theme is recomposed so that Emma's narrative ends in a different key than it began. This instance of directional tonality is a reflection of Emma's personal development, as she evolves from foolish single girl to wise(ish) married woman. She *settles down*, narratively and tonally speaking. Embedded within Emma's directional arc, we can trace a functional path to her trajectory. Table 10 shows the keys of the "Main" theme at the pivotal moments in the narrative: the beginning, the climax, and the ending. Taking the final key of G major as Emma's home key (since this is where she settles), we can assign the Roman numerals II–V–I to this progression of keys/events: II (A major) is where she agitatedly begins, V (D major) is where she attains denouement, and I (G major) is where she finds resolution. Thus a functional P–D–T (pre-dominant, dominant, tonic) cadential progression adumbrates her character arc.

Unlike Anne of *Persuasion* and Benjamin of *The Graduate*, Emma's problem is *too much* initiative and activity; thus the directional tonality of her story moves in the opposite direction—downward rather than upward.

FANTASTIC MR. FOX (2009)

Elements featured: associative tonality, transposition, tonal wink, tonal agency, intertextual tonality, singular key, parallel relationship, directional tonality, tragic-to-triumphant arc, functional tonality.

RELEVANT PLOT SYNOPSIS

Mr. Fox is restless in his sedate life as a husband and father. Immature and selfish, he longs for the excitement and danger of his youth, when he was a daring, dashing burglar. He makes the foolhardy decision to buy a tree house (rather than living below-ground, as foxes normally do) directly adjacent to the farms of Boggis, Bunce, and Bean. The realtor warns Mr. Fox about the three malevolent farmers, but Mr. Fox is undeterred.

Mrs. Fox's nephew, Kristofferson, comes to live with them, and Ash takes a violent dislike to his cousin, who is superior to him in every way, and whose great athleticism wins Mr. Fox's effusive approval—approval that awkward, oddball Ash desperately seeks.

Mr. Fox begins stealing from the Boggis, Bunce, and Bean farms, becoming increasingly reckless. The angry farmers set out to hunt and kill Mr. Fox. They manage to shoot off his tail (which Mr. Bean then wears as a necktie), but Mr. Fox escapes. The farmers demolish the entire hill on which his tree house was located, in an attempt to dig him out, only to discover that the Foxes have dug an escape tunnel. The farmers encamp by the tunnel entrance and wait for the Foxes to be driven to the surface for food. Underground, Mr. Fox encounters all the other local animals whose homes were destroyed as a result of his actions. As they begin to face starvation, Mr. Fox leads a digging expedition to tunnel into the three farms. They burgle food for the group and hold a feast, at which Mr. Fox toasts and boasts with great hubris. Ash and Kristofferson reconcile in order to plan a mission to retrieve Mr. Fox's tail from Bean's farm. They are interrupted by the arrival of Bean's wife, and Ash escapes while Kristofferson is captured.

The farmers flood the tunnel, and the animals are forced into the sewers. Mr. Fox confronts Rat (Bean's security guard), who informs him that the farmers are holding Kristofferson hostage. Mr. Fox and Rat fight, and as Rat lays dying, he grudgingly divulges Kristofferson's location. Mr. Fox requests a parley with the farmers in the town square, where he will surrender in exchange for Kristofferson's release. The farmers have planned an ambush, but the animals anticipate it and launch a counterattack. During the commotion, Mr. Fox and Ash slip into Bean's farm, where Ash finally proves himself to his father by freeing Kristofferson.

The animals grow comfortable living in the sewer system, and it becomes their new home. One day, Mr. Fox leads them through a tunnel that happens to lead directly into the Boggis, Bunce, and Bean International Supermarket. Elated to have discovered this infinite food source, they dance.

The tonal design of this charming, animated Wes Anderson film owes its foundation to the main title theme (composed by Ennio Morricone) of Sergio Leone's

FIGURE 18. Basic melodic content of "BBB" theme, which is used to generate six slightly different versions.

iconic Spaghetti Western, *For a Few Dollars More* (1965). Anderson's showdown between farmer and fox in this film is a parody of the classic cowboy-adversary showdown, and Alexandre Desplat based his showdown theme on Morricone's "For a Few Dollars More" theme (see Video 11 ✱). The basic melody of Desplat's theme is shown in Figure 18, and I refer to this core melodic material as the "BBB" theme (where BBB stands for Boggis, Bunce, and Bean). The BBB theme is used as the basis for six distinct cues in this film:

"Great Harrowsford Square"
"Stunt Expo 2004"
"Boggis, Bunce, and Bean"
"Bean's Secret Cider Cellar"
"Just Another Dead Rat in a Garbage Pail"
"Fox Problems"[14]

Like the iconic Morricone theme, the BBB theme features a twanging mouth harp, a crystalline whistled melody, monosyllabic words grunted rhythmically by a chorus, and the key of D minor. In fact, while the D-minor BBB theme "Great Harrowsford Square" is playing during the crucial climax of the narrative, the closed captions read: "For a Few Dollars More playing." Clearly, the Morricone theme begat the Desplat theme, and its key of D minor is employed to intertextually allude to the Spaghetti Western melodrama in *Fantastic Mr. Fox*.

D minor is first introduced in Badger's real estate office, when he plays a recording of the children's rhyme "Boggis, Bunce, and Bean" (one of the BBB themes listed above) for Mr. Fox, to warn him against buying the tree adjacent to their farms (0:09:18). The BBB theme was heard only moments earlier in the key of B♭ minor (when Mr. Fox first laid eyes on the BBB farms) (0:07:26), but now the theme is in D minor and it has lyrics attached—lyrics so central to the narrative that they were actually featured on the very first frame of the film, before the music or narrative even begin:

---

14. I assigned the name "Fox Problems" because this theme is not listed or credited anywhere. The other five titles are listed in the official soundtrack album.

> BOGGIS AND BUNCE AND BEAN
> ONE FAT,
> ONE SHORT,
> ONE LEAN,
> THESE HORRIBLE CROOKS
> SO DIFFERENT IN LOOKS
> WERE NONE THE LESS EQUALLY MEAN.

This D-minor music is also diegetic, meaning that Mr. Fox is actually hearing it (as opposed to the aforementioned B♭-minor iteration of the BBB theme, which was non-diegetic and heard only by *us*). This marked presentation of D minor draws attention to itself as Mr. Fox's nemesis key (his Spaghetti Western villain) and foreshadows the battle royal (culminating in a classic showdown in the town square) in which Mr. Fox will entangle himself. And apart from his showdown with Boggis, Bunce, and Bean, Mr. Fox also encounters his *other* nemesis—the lone black wolf—in the key of D minor (the "Canis Lupus" theme) (1:14:36).

Complementing the "nemesis" status of D minor is its parallel major, which is Mr. Fox's identity key. D major is associated with Mr. Fox's character—his swashbuckling joie de vivre, his ambitious aspirations, and his desire to be "fantastic" in every situation in life. To begin with, the narrative opens with a D-major theme branded with Mr. Fox's identity (titled "Mr. Fox in the Fields"), as we are acquainted with his character and life (0:01:57). The next D-major cue (Burl Ives's "Fooba Wooba John") acquaints us with his aspiration (to live aboveground) (0:04:56), and the next cue (the "Mr. Fox in the Fields" theme) shows him achieving his aspiration (viewing his dream tree house with a realtor) (0:06:36). Even Mr. Fox's ubiquitous whistle aptly represents his identity ("That's my trademark"), consisting of the D-major-centric pitches D-A-D. The film is peppered with his D-major whistle, as though Mr. Fox is scent-marking the soundtrack with his identity key, in a subtle and crafty form of tonal agency.

Mr. Fox also employs tonal agency when he diegetically produces D-major music in his grandest moment. His toast at the animals' subterranean banquet dinner had been interrupted by a disaster (the cider flood), which led to his turning point of self-realization (recognizing that his actions had endangered the group). Before rallying the animals for a second offensive against the farmers-in-arms, Mr. Fox mentions that he was dissatisfied with his previous toast and thus resets the situation ("I'm gonna start over"). He reaches down to turn on his "WALK-SONIC" transistor radio, playing Georges Delerue's "Le Grand Choral" D major (0:58:48) and commences with his *new* toast, which is an inspirational speech. Now in his D-major element, Mr. Fox rallies the animals to glorious victory with his charismatic leadership. And in a marvelous instance of both functional tonality and tonal agency, the cue that accompanied Mr. Fox's *first* toast was in A major

(Delerue's "Adagio") (0:49:23), which acted as a dominant springboard into the D major of his second toast (Delerue's "Le Grand Choral") (see Video 12 ✶).[15]

The key of B♭ minor belongs to the evil farmers Boggis, Bunce, and Bean—and knowing Anderson's quippy style, the use of B♭ for Boggis, Bunce, and Bean is most likely meant as an alliterative tonal wink. This key is used sparingly and always with the purpose of illustrating the cumulative horribleness of this malevolent trio (not the individual farmers, but the three together as a unit). B♭ minor first occurs when Mr. Fox lays eyes on their three farms adjacent to the tree house (briefly mentioned above) and soon proceeds to accompany Badger's description of each of the three farmers. Its final iteration occurs when Boggis, Bunce, and Bean pool together their 108 workers to fight against the animals. Each of the three instances of B♭ minor involves the "Boggis, Bunce, and Bean" theme, which is one of the BBB themes; but since the BBB theme can occur in nine different keys (see Table 11), the key of B♭ minor *can* be separated from its thematic attachment to function independently as an associative key. Incidentally, it's worth noticing that the Boggis Bunce Bean key of B♭ minor is the *hexatonic pole* of Mr. Fox's key of D major, a harmonic relationship loaded with dark meaning made famous by nineteenth-century works such as Wagner's *Parsifal*.[16] (And Wes Anderson will go on to reuse the key of B♭ minor as the villain key in *The Grand Budapest Hotel*, five years later, as we see in chapter 4.)

G major is the key of redemption in this film. Several preexisting and original cues operate in this key, and they are mainly used to accompany key moments in which characters redeem themselves. Ash, after making Kristofferson cry, overcomes his petulant jealousy and turns on his model train set (which plays Desplat's G-major theme "High-Speed French Train") to console him (0:14:13). Georges Delerue's "Une petite île" in G major underscores Mr. Fox's apology to Mrs. Fox, when he admits that his pride put them all in danger, and he promises to change (0:51:55). The Beach Boys' G-major "I Get Around" accompanies Mr. Fox as he formulates his *new* plan to fix the mess he created with his *old* plan, by harnessing each animal's strengths, rather than relying solely on his own (1:00:10). And finally, the animals dance joyously to the Bobby Fuller Four's "Let Her Dance" in G major when Mr. Fox has redeemed himself as a leader by (adventitiously) leading them into the Boggis, Bunce, and Bean International Supermarket, where they are surrounded by more than enough food to ensure their survival (1:20:01). This moment also redeems Mr. Fox as a father, as he and Ash finally share a significant exchange of mutual respect:

---

15. As Winters (2012) and McQuiston (2017: 477) have noted, *general* musical agency (characters playing music) is a trope in several Wes Anderson films.

16. For seminal writings on the hexatonic pole, see Cohn (2004, 2006, and 2012).

TABLE 11 BBB-based themes and their key areas in *Fantastic Mr. Fox*

Cue	Key
"Great Harrowsford Square"	C minor (→ D minor → F minor)
"Stunt Expo 2004"	D minor (→ F minor)
"Boggis, Bunce, and Bean"	B♭ minor (→ D♭ minor → F♭ minor → A♭♭ minor → B♭ minor)
"Just Another Dead Rat in a Garbage Pail"	C minor (→ G minor)
"Bean's Secret Cider Cellar"	C minor
"Fox Problems"	A minor

Mr. Fox: "To our . . . *survival*. How was that?"
Ash: "That was a good toast."

With close-up shots of their meaningful gaze, their relationship is now implied to be mended, and Ash turns on the "WALK-SONIC" (now worn on *his* hip) to play "Let Her Dance" in G major. And interestingly enough, one of the film's malefactors, Rat, *also* redeems himself, by revealing where Kristofferson is being held, just seconds before he dies. But Rat's sad moment of not-quite-pure-hearted redemption-too-late (he admits he never would have helped if he hadn't been dying) is in G *minor* instead of G major (with Desplat's theme "Just Another Dead Rat in a Garbage Pail"—the only moment of G minor in the film) (0:57:21).

The key of C minor accompanies the major confrontations in this film: Mr. Fox confronts Rat (0:25:50), Rat confronts Mrs. Fox (0:56:12), the farmers present their ransom note to Mr. Fox (0:55:11), and Mr. Fox presents his ransom note to the farmers (1:01:32). Both C-minor themes used in these situations are based on the BBB theme, but, as mentioned earlier in the context of B♭ minor, C minor was chosen for these confrontational scenes out of nine possible keys.

This film soundtrack makes ample use of key transposition. The BBB theme is transposed to several different keys in the context of six themes—and even within these themes, the BBB theme sees further transposition via modulation (Table 11). These modulatory sections are modular enough to allow the BBB theme to be excerpted as a standalone theme in nine different keys. For example, in the great "high-noon"-style standoff in the town square, the BBB theme consists of the D-minor middle section of the "Great Harrowsford Square" theme (1:02:28) (excluding its C-minor opening and F-minor closing sections).

Another example of transposition in this film involves the "Mr. Fox in the Fields" theme, which is transposed from the key of Mr. Fox's "fantastic" identity (D major) to C major when Mr. Fox concocts a not-so-fantastic plan that ends disastrously, trapping all the animals in dire straits (0:41:05). Conversely, the preexisting work "Petey's Song" (by Jarvis Cocker) is transposed *to* D major during the closing credits (1:22:30)—from its original E♭ major, which was heard during the campfire montage (0:42:55)—to close out the film in the context of Mr. Fox's character key. The meticulous use of sound effects in this film is discussed in chapter 5.

The grand finale showdown of this film traces a tragic-to-triumphant arc, beginning in a tense D minor (of the "Great Harrowsford Square" theme) and ending in a jubilant D major (of Georges Delerue's "Le Grand Choral") once the animals have achieved success. This tragic-to-triumphant arc is embedded within a larger arc of directional tonality, as the film's soundtrack is not tonally closed but begins in E major and ends in D major. Since D major is Mr. Fox's identity key, and the key of the animals' Fox-led triumph over the farmers, we can see in the soundtrack's overall tonal trajectory a directional reflection of Mr. Fox's *personal* trajectory, as he matures into his best self by the end of the film. Like the protagonist of *Emma*, Mr. Fox must settle down in order to grow up.

CLOSING THOUGHTS

The seven films discussed in this chapter wouldn't ordinarily find themselves grouped together, but through the lens of film tonality, we can draw a palpable connection. The directional tonality in each of these soundtracks traces the trajectory of the protagonists, as they rise up or settle down to achieve their happy endings. And the use of transposition further affirms directionality by moving a single musical theme through multiple keys, inviting us to notice the tonal evolution of that theme.

Now on to chapter 4, where we explore the last remaining technique (tonal symmetry) we have not yet encountered in analysis, and where we analyze two films in their entirety, rather than looking at selected excerpts.

4

A Tale of Two (Tonally Symmetrical) Films

In chapters 2 and 3 we surveyed a number of film excerpts to gain a breadth of understanding for how tonal techniques can function in the soundtrack context. In this chapter we go into greater depth by analyzing two films in their entirety—*The Talented Mr. Ripley* (1999) and *The Grand Budapest Hotel* (2014)—to see how a film's overarching tonal design can enable the soundtrack to function as a cohesive entity. I have paired these two films together because both are built around a flashback, and the flashback structures the tonal design in a way that reflects the central theme and symmetry of the narrative. Comparison of these two soundtracks allows us to draw some intriguing parallels and gain insights about how a tonal design can articulate a film's narrative structure.

THE TALENTED MR. RIPLEY (1999)

Techniques featured: associative tonality, tonal coda, transposition, singular key, tonal agency, intertextual tonality, tonal pairing, relative relationship, parallel relationship, cadential frustration, tonal symmetry.

RELEVANT PLOT SYNOPSIS

Tom Ripley (Matt Damon) is a downstairs member of 1950s Manhattan high society, who longs to join the upstairs echelon. When he is mistaken for a Princeton alumnus by wealthy socialite Mr. Herbert Greenleaf (James Rebhorn), Tom takes the opportunity to forge a new path for himself. He pretends to have known Mr. Greenleaf's son Dickie (Jude Law) at Princeton, and Greenleaf sends Tom to Italy to convince his prodigal son to return home.

Once in Italy, Tom forges a spurious connection with Dickie and his girlfriend Marge (Gwyneth Paltrow) by manufacturing a fake past and feigning to share the same interests. Tom becomes obsessed with Dickie and falls in love with him, though Dickie does not return his affection. When Dickie's old friend Freddie (Philip Seymour Hoffman) arrives in town, Dickie instantly tires of Tom and discards him. Tom tries unsuccessfully to cling to Dickie, and he ends up murdering Dickie in a jealous rage. After disposing of his beloved's body, Tom takes over Dickie's identity and uses his money to create a posh new life for himself.

Tom artfully juggles both identities until Freddie becomes suspicious about Dickie's absence, and Tom murders Freddie to cover his tracks. When Freddie's and Dickie's bodies are discovered, Tom comes under police investigation. To escape the suspicions, Tom sets sail on a cruise with his new lover, Peter (Jack Davenport).

Onboard the ship, Tom is entrapped in his double role when he encounters Meredith (Cate Blanchett), a mutual friend of Peter's who knows him as "Dickie." Tom realizes he cannot prevent Meredith and Peter from discovering his dual identity and deception. He cannot resolve the situation by murdering Meredith because she is accompanied by her family, so he is forced instead to murder Peter.

The Talented Mr. Ripley might be considered Anthony Minghella's greatest film, for the complexity and depth of its main protagonist, Tom Ripley, and the marvelously conflicting feelings he engenders in the audience (it's one of those unusual films where you find yourself rooting for the villain rather than his victims). This film masterfully inveigles the viewer-listener into wishing Tom success in his murderous endeavors and sympathizing with how unfair the world is for this charismatic sociopath.

The soundtrack for this film likewise sympathizes exclusively with Tom, both in its preexisting music and original scoring: the preexisting works reflect Tom's musical interests with classical works he loves passionately and jazz numbers he uses to achieve his goal (of infiltrating Dickie's world).[1] And unlike most films, in which originally composed themes are used to designate a variety of characters, places, and plot elements, *all* of the themes in *The Talented Mr. Ripley* are written solely for Tom—to reflect the various emotional and psychological stages of his experience. According to Heather Laing, nondiegetic film music has historically been used to represent the interiority of female characters, with the main gender exception being male characters who are "emotionally or psychologically disturbed" (2007: 139). Applying Laing's findings to the Rick character in *Casablanca* (1942), David Neumeyer states that "music burdens the characters it underscores because its use suggests that the character cannot or will not adequately express himself" (2015: 151). Likewise in *The Talented Mr. Ripley*, musical interiority is employed solely for

---

1. For in-depth discussions on the role of classical and jazz music in both defining and masking sexuality in this film, see Decker (2012) and Poluyko (2011).

Tom, because he is so heavily burdened with dark secrets that *all* of the underscore must serve to unburden him. Furthermore, with specific reference to male characters who are *musicians* (performers or composers), Laing notes that "these men are ultimately unable to find an unproblematic place in society, or even in life" (2007: 140), which is certainly an apt description of Tom Ripley's struggles.

These nine themes (✶) earned composer Gabriel Yared an Academy Award nomination for Best Original Score. Seven of the nine themes are in D minor, and we are aurally immersed in this key even before we visually encounter the film. The film begins with D-minor music ("Lullaby for Cain" theme) on a black screen, and thirty seconds later Tom is revealed, sitting in his ship cabin, deep in thought after having just murdered Peter. Tom's opening line—"If I could only go back"—takes us inside his head as he recounts the whole chain of events that led him to his current state. The entire film takes place inside Tom Ripley's mind, as one continuous flashback. After this mental replay the film ends with a mirror inversion of the film's beginning: a closing shot of Tom sitting on the bed, contemplating Peter's murder, accompanied by D-minor music ("Death" theme), which dissolves into a black screen. The opening and closing of the film are thus symmetrically framed and create a palindromic structure for the film (✶). The film begins and ends in the same time and place, and the key of D minor corroborates this narrative arc and symmetrical film structure.

The film associates D minor with Tom's darkness, deception, and despair. After the opening theme ("Lullaby for Cain"), the first few minutes of the film are devoid of music, until another D-minor theme ("Mischief") marks Tom's first lie (lying to Mr. Greenleaf that he knew his son Dickie at Princeton)—the lie that sets off his elaborate chain of deception. The next instance of D minor occurs when Tom falsely introduces himself to Meredith; then again when he falsely introduces himself to Dickie and Marge; when Tom tricks Dickie into friendship; when Tom murders Dickie; when Tom assumes the identity of Dickie Greenleaf; when Tom murders Freddie; when Tom lies to the police—the list goes on and on. All of Tom's darkest and most deceptive moments are underscored by the seven D-minor themes. D minor occurs in forty-six occasions throughout the film, and this key accounts for more than half of the film's total musical time (Table 12).

Table 13 shows all forty-six cues in D minor, forty-five of which are nondiegetic original themes. Notably, only one D-minor cue in this film is diegetic, and this preexisting work dramatically reveals *why* Tom's tragic themes are all in the key of D minor.

Days after Tom has murdered his beloved Dickie, he attends a performance of Tchaikovsky's *Eugene Onegin*. In the film the opera picks up in the midst of the pistol duel scene between Onegin and Lenski (Act II Scene 2). Tom watches with rapt attention as Onegin murders his dearest friend, and when he and Zaretski pronounce Lenski dead ("Killed?" "Killed!"), they do so in the key of D minor (see

TABLE 12 Allotment of musical time* per key in *The Talented Mr. Ripley*

Key	Minutes	Percentage of Film's Total Musical Time (%)
D	46.23	55
F	10.17	12
E♭	6.68	8
B♭	5.05	6
other keys	16.34	19

* Total amount of time when all the musical "silence" (in which no music is present) is stripped away

Video 13 ✶). The scene begins in B minor, and D minor is established precisely at the instant Lenski is confirmed to be dead, a moment that corresponds with an extreme visual close-up of Tom's horror-stricken, tear-streaked face, as he sees the reflection of his own dark deeds in the opera's murder scene (Figure 19). This close-up suggests that Tom perceives a parallel between his murder of Dickie and Onegin's murder of Lenski, and shares the vicarious anguish of the operatic character. The key of D minor imprints on Tom at this dramatic moment, and his tragedy becomes powerfully and permanently associated with this key. Because the film's *plot* begins at the end of the *story*, everything we witness in this film is being filtered to us through Tom's memory.[2] This impactful operatic moment recasts his memory in the key of D minor. The presence of D minor in Tom's recollection emanates from this pivotal Tchaikovsky excerpt, which incidentally occurs at the precise midpoint of the film. This D-minor kernel unfurls outward from the heart of Tom's story and colors his perception of his entire devastating tale.

In the harmonic landscape of this film the most prominent tonal relationship is the one between D minor and F major. These relative keys underscore the dichotomy of Tom's polar halves: his deceptive, dark side and his sanguine, cheerful side. The relative major of Tom's tragic key, F major is used to represent the narrative antipode of the D-minor associations, and it underscores the few fleeting moments of happiness and hope in Tom's troubled life. Given the dark nature of Tom's story, F major does not appear often—but that makes it all the more marked. The first instance of F major occurs in the concert hall where Tom works (as a restroom attendant), just after having met Mr. Greenleaf: after being coldly excluded from the chamber music concert (and the upper-class world he longs to inhabit), Tom appropriates the concert-hall stage after hours and plays Bach's exuberant F-major *Italian Concerto* on the grand piano (0:04:08).

---

2. "Story" is the chronological sequence of events as they occur in real time, while "plot" is the order of events presented in a film, not beholden to actual chronology.

TABLE 13 D-minor cues (in the order in which they appear) in *The Talented Mr. Ripley*

D-minor Cue	Status	Type	Timestamp
"Lullaby for Cain"	nondiegetic	original	0:00:00
"Mischief"	nondiegetic	original	0:02:29
"Mischief"	nondiegetic	original	0:08:58
"Mischief Mobilized"	nondiegetic	original	0:11:17
"Mischief Mobilized"	nondiegetic	original	0:14:07
"Mischief"	nondiegetic	original	0:16:22
"Mischief Mobilized"	nondiegetic	original	0:18:18
"Proust"	nondiegetic	original	0:23:58
"Proust"	nondiegetic	original	0:38:33
"Syncopes"	nondiegetic	original	0:39:39
"Proust"	nondiegetic	original	0:41:58
"Death"	nondiegetic	original	0:42:48
"Syncopes"	nondiegetic	original	0:48:27
"Proust"	nondiegetic	original	0:49:38
"Proust"	nondiegetic	original	0:54:18
"Death"	nondiegetic	original	0:57:29
"Italia Corrupted"	nondiegetic	original	0:58:08
"Syncopes"	nondiegetic	original	1:01:25
"Mischief Mobilized"	nondiegetic	original	1:03:47
"Mischief Mobilized"	nondiegetic	original	1:06:06
thematic medley	nondiegetic	original	1:07:18
Tchaikovsky *Eugene Onegin*	**DIEGETIC**	**PREEXISTING**	**1:10:08**
"Proust"	nondiegetic	original	1:11:40
"Promise"	nondiegetic	original	1:14:44
"Mischief Mobilized"	nondiegetic	original	1:15:41
"Proust"	nondiegetic	original	1:16:35
"Mischief Mobilized"	nondiegetic	original	1:18:26
"Proust"	nondiegetic	original	1:21:13
"Proust"	nondiegetic	original	1:23:40
"Proust"	nondiegetic	original	1:26:19
"Proust"	nondiegetic	original	1:29:16
"Proust"	nondiegetic	original	1:34:24
"Syncopes"	nondiegetic	original	1:37:41
"Italia Corrupted"	nondiegetic	original	1:38:09
"Italia Corrupted"	nondiegetic	original	1:53:20
"Death"	nondiegetic	original	1:55:24
"Proust"	nondiegetic	original	1:56:05
"Italia Corrupted"	nondiegetic	original	1:56:48
"Death"	nondiegetic	original	1:59:24
sustained D pitch	nondiegetic	original	2:01:14
"Mischief"	nondiegetic	original	2:03:19
"Death"	nondiegetic	original	2:08:31
thematic medley	nondiegetic	original	2:10:26
"Death"	nondiegetic	original	2:13:15
"Italia Corrupted"	nondiegetic	original	2:14:01
"Death"	nondiegetic	original	2:18:18

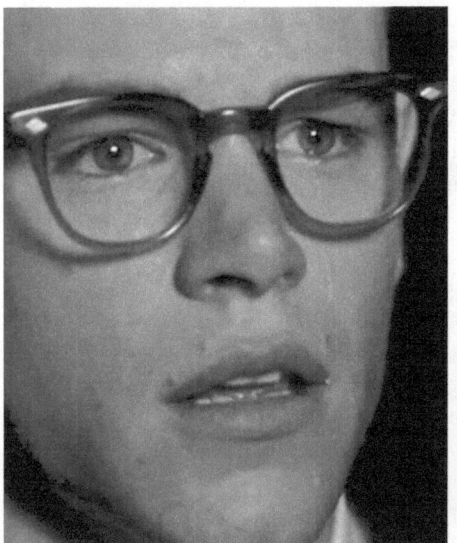

FIGURE 19. Tom's face (left) as he watches Onegin holding his dead beloved on the opera stage (upper right), with the extreme close-up of his face implying that he realizes the parallel to his own murderous deed (lower right).

The choice of Bach's "Italian" Concerto is an allusion to Tom's impending trip to Italy, which Mr. Greenleaf commissions in the very next scene. F major occurs again (in the form of the "Italia" theme) when Tom arrives in the lovely Italian seaside village, filled with blithe optimism for his new life (0:10:05). The third instance of F major (the "Italia" theme again) occurs when Tom is out sailing with Dickie and Marge, basking in the radiant warmth of Dickie's fickle attention (0:25:35). Tom is overflowing with sanguine joy as Dickie enthusiastically makes a string of plans to include Tom in his life. We briefly hear F major a fourth time (the folk tune "Roma," played by a street accordionist) when Tom is enjoying Dickie's company at a café in Rome (0:33:54). This brief happiness is quickly dissolved when Freddie shows up at the café and whisks Dickie away from Tom. Freddie's magnificently arrogant entrance in his stunning red Alfa Romeo convertible snuffs out this bit of F major, as he instantly commandeers Dickie's attention away from Tom. This is the beginning of the end of Tom's happiness, and F major is absent from the narrative for quite some time, as Dickie tires of Tom, Tom frantically senses it, and eventually murders Dickie.

The final iteration of F major (Bach's *Italian Concerto*) occurs much later on, during a montage sequence illustrating how Tom takes over Dickie's charmed life. The Bach begins nondiegetically at Café Dinelli (1:19:32), when Tom realizes that he has succeeded in hoodwinking everyone (Marge, Peter, and Meredith) about Dickie's disappearance, and follows him as he tours a grandly furnished apartment. Now having rented that apartment (using Dickie's money, of course), Tom enjoys a cozy Christmas in luxurious style.[3] Finally, the montage leads to a spectacular Alfonsi piano Tom has purchased for his apartment, and it is now revealed that *Tom* was the one playing the Bach this whole time. With this long F-major sound advance, it is as if Tom figuratively stretched out his fingers to grasp onto the piano keys and used them to pull himself all the way from Café Dinelli (after settling Marge and Peter's questions) into the precise situation he had long dreamed of: independence, wealth, and culture. Tom's tonal agency in stealthily producing this F-major work as a sound advance allows him to masterfully manipulate everyone and everything to his advantage. And of course, the choice of this "Italian" concerto reflects Tom's celebration of his bright future here in Italy. But this return to the *Italian Concerto* puts Tom right back where he started: chasing after a happiness he is not destined to keep. He is interrupted (on a dramatic dominant chord) by Freddie pounding on his apartment door. The jig is symbolically up as Freddie obstructs Tom's attempt to cadence in F major (more on this later).

After this the optimism of F major is meaningfully deflated to a dejected F minor. Tom endures a grueling interview at the police station, then plays the doleful first movement of Vivaldi's *Stabat Mater* in F minor on the piano in Peter's apartment (1:44:53). While playing, Tom cryptically alludes to his crushing burden of secrets, as he realizes sadly there is no way to salvage hope or happiness. Immediately following this, Peter conducts an orchestral performance of the *Stabat Mater* from the high balcony of a melancholy cathedral (1:47:11). The text of this musical passage depicts a weeping Mary standing at the foot of the cross where her Son hangs:

Stabat mater dolorosa	The sorrowful mother stood
Juxta crucem lacrymosa	Before the cross, weeping
Dum pendebat Filius.	Where hung her Son.

These lines put Tom in the position of the mourner and Peter in the position of the martyr (he does, after all, die for Tom's sins). The visuals in this shot (✶) depict Peter as the very embodiment of purity and innocence—positioned on high in the stratospheric balcony (and flanked by an angelic boy alto, not shown in this shot)—as a heartbroken Tom gazes up and realizes he will never attain redemption or happiness. The sounding of the parallel minor key corroborates the crushing of Tom's last shred of hope—with no return to F major.

The key of E♭ minor in this film is associated with the loneliness of unrequited romantic longing, and its parallel E♭ major accompanies moments of romantic union. The first instance of E♭ major occurs at the end of a montage sequence in which Tom pursues Dickie's love (see Video 14 ✶). The song "My Funny Valentine" begins nondi-

---

3. During this Christmas scene Yared builds a second layer of F-major music atop the *Italian Concerto*: a boys' choir singing a Latin text with organ accompaniment.

egetically in a disconsolate minor key, while Tom jealously watches Dickie kiss Marge (0:28:42). In the next scene of the montage Tom gathers detailed information about how Marge won Dickie's heart, as she recounts how she and Dickie fell in love over this song. Tom rides off with Dickie on a moped in the next scene, without Marge. Finally, the montage ends in a Naples jazz club where it is revealed that *Tom* is (diegetically) singing the song, with Dickie on saxophone. He ends the song in the key of E♭ major after he succeeds in getting Dickie alone in what *he* fantasizes to be a romantic situation. The two of them are shown alone in Dickie's apartment following their jazz club performance, where Dickie passionately kisses his new icebox (which Tom bought for him) and exclaims erotically, "*Ohhhh*, I could *fuck* this new icebox, I love it so much." His orgasmic "*Ohhhh*" is sound-advanced while we are still visually present in the jazz club, and at the same time, the brushed shimmer of the hi-hat spills over from the jazz club into Dickie's apartment as a sound lag.

The simultaneous sound lag and sound advance merge these two scenes together. Remembering that this entire narrative takes place not in reality but inside Tom's perception of reality, this dual-directional sound bridge is Tom's way of connecting the erotic passion of his androgynous, sexually charged performance (of "My Funny Valentine") with Dickie's titillating outburst at home afterward.[4] The narrative arc of this montage thus begins with Dickie kissing Marge and ends with Dickie (symbolically) kissing Tom. Immediately after Dickie's orgasmic gesture, the camera reveals Tom smiling radiantly and smoking the proverbial postcoital cigarette. (Tom doesn't smoke ordinarily, so it holds marked significance in this context.) Tom uses this song to covertly couple himself with Dickie, and the key of E♭ major underscores his perception of his romantic triumph. Much like the F-major sound advance discussed earlier, Tom uses this sound advance to furtively bootstrap his way into the situation he desires. Tom's tonal agency with the key of E♭ major is further accentuated by the fact that the *original* version of "My Funny Valentine" (sung by Chet Baker) ended in C minor rather than E♭ major (with the final vocal pitch of E♭ serving as 3̂, rather than 1̂).[5] Tom's version thus purposefully recomposes the song to end in his "romantic union" key of E♭ major. This recomposition is powerfully symbolic of Tom's rose-tinted reinterpretation of events, and his tonal agency in using music to get what he wants.

Tom uses the key of E♭ major to connect himself with Dickie in a second instance. After being abandoned by Dickie (upon Freddie's arrival), Tom embarks on a lonely sightseeing excursion in Rome, accompanied by the despondent

---

4. When Tom first hears "My Funny Valentine" in his apartment in New York, he remarks, "I don't even know if this is a man or a woman." This androgyny has a primal appeal for Tom, as we observe him effeminize himself in several scenes.

5. Chet Baker's C-minor "My Funny Valentine" does flirt with the relative major at several points in the song but ends emphatically in C minor.

FIGURE 20. Tom passes a pair of romantic men and then passes a pair of celibate nuns, during his lonely E♭-minor excursion.

E♭-minor "Promise" theme (0:36:49). During this E♭-minor montage Tom misses Dickie acutely as he looks longingly at two young men romantically canoodling in the street. The next shot immediately replaces this amorous couple with a pair of chaste nuns—the epitome of romantic singleness and celibacy—further accentuating what Tom *wants* versus what he *has* (Figure 20).

Tom exerts his tonal agency to transform this gloomy E♭ minor into a gleeful E♭ major by supplanting the nondiegetic "Promise" theme with his own diegetic performance of Bing Crosby's "May I?" (0:37:26) (see Video 15 ✶). This song transports

FIGURE 21. Dickie appears when Tom performs "May I?" in E♭ major, and they are reunited (in the mirror). They even seem visually paired, with their matching black-and-white outfits and blonde hair (perhaps in anticipation of Tom assuming Dickie's identity).

Tom away from the dark lonesome night and into the warm glow of Dickie's bedroom, where he (Tom) dresses up in Dickie's clothing and dances before the mirror in a sexually suggestive manner. By starting in on the same tonic and transforming the mode, he turns his E♭-minor rejection (after being deserted by Dickie) into an E♭-major reunion (as he dons Dickie's attire in an intimately invasive way). And, incidentally, the moment Tom performs this song-and-dance, Dickie returns to him (Figure 21), thus granting Tom's E♭-major performance an even greater sense of agency.

The film ends with a reversal of this parallel mode transformation, as E♭ major becomes E♭ minor when Tom is plunged back into romantic loneliness. Tom and Peter joyously sing Gilbert & Sullivan's "We're Called Gondolieri" (in E♭ major) from the bow of the ship, as they sail into the sunset (2:05:09).[6] Tom is celebrating the fact that, like the young gondolier in Gilbert & Sullivan's comic opera *The Gondoliers*, he has mistakenly been identified as the king's son and heir (Mr. Greenleaf has decided to bequeath Dickie's fortune to Tom). But the mirror inversion of the earlier E♭ minor-to-major, nondiegetic-to-diegetic transformation has now turned into an E♭ major-to-minor, diegetic-to-nondiegetic transformation (Figure 22). With this reverse shift, Tom's window of romantic happiness closes (and he is reluctantly forced to murder Peter a few minutes later).

6. Tom's rendition of "We're Called Gondolieri" begins in E major and dips down slightly to end in E♭ major. Like his "My Funny Valentine," this song begins in one key but ends in his "romantic union" key of E♭ major.

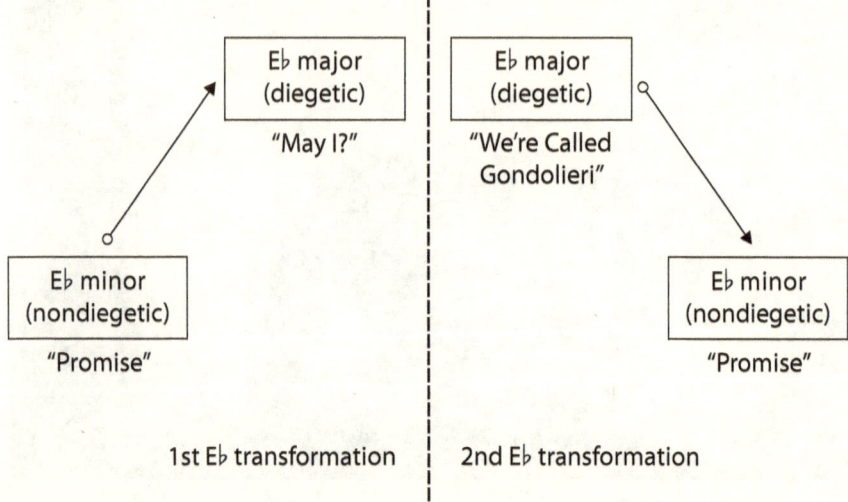

FIGURE 22. Mirror-inversion E♭ transformations of mode (major/minor) and type (nondiegetic/diegetic).

The key of B♭ major accompanies Tom's jealous covetousness for that which he cannot attain—namely, acceptance into aristocratic society. With unmistakable yearning on his face, Tom furtively watches a performance of the second movement of Beethoven's Piano Quartet, Op. 16 (in B♭ major) from behind the thick velvet curtain, before being scornfully rebuffed by an audience member (0:03:20). Later, in his squalid little apartment, Tom assiduously studies a recording of Charlie Parker's "Ko-Ko" (in B♭ major) in an attempt to educate himself in the jazz idiom, to feign sophisticated taste among the cultural elite (0:06:09). "Mache dich, mein Herze, rein" from Bach's *St. Matthew Passion* (in B♭ major) accompanies Tom as he pores over a Social Register in which Dickie's blueblood lineage is illustrated (0:06:33).[7]

And finally, Tom watches Dickie and Freddie with sullen jealousy as the two huddle intimately in a record store listening booth, listening to Sonny Rollins's "Tenor Madness" (in B♭ major) (0:34:34). This music has the effect of instantly alienating Tom from Dickie, both emotionally and physically. "Tenor Madness" begins nondiegetically at the moment Freddie makes his grand arrival in the red convertible and diverts Dickie's attention. "Tenor Madness" is diegeticized in the next scene, when Dickie and Freddie are shown listening to it in the record store, leaving Tom standing outside, alone. Thus what began as emotional isolation is transformed into *physical* isolation, when the music is diegeticized. Recall that the music playing before

---

7. A Social Register (the American equivalent of a British peerage book) is a publication that lists members of the social elite, historically focused on old-money families of the Northeast.

Freddie's entrance was (the folk tune "Roma") in F major, so Freddie's unwelcome arrival supplants Tom's F-major happiness with B♭-major jealousy.

In addition to the prominent key associations in this film's tonal design, there are several significant instances in which themes are transposed from their original keys to new keys for narrative reasons. The sunny "Italia" theme is transposed from F major to the tragic key of D minor (which I have labeled as "Italia Corrupted")[8] when Tom hides Dickie's corpse (0:58:08), when Tom forges a suicide note by Dickie to cover up his murder (1:38:09), and when Tom lies to Marge about Dickie's disappearance (1:56:48). Another transposed theme is the "Promise" theme, which normally appears in the key of romantic loneliness (E♭ minor), but which is transposed to the duplicitous key of D minor when Tom jilts the lovelorn Meredith (after the opera in Rome) by lying about his involvement with Marge (1:14:44).[9] Both of these themes ("Italia" and "Promise") acquire additional nuance when they are transposed to D minor from their original keys. In the case of the "Promise" theme, the amalgamation between E♭ minor (romantic loneliness) and D minor (duplicity) superimposes Tom's deception with Meredith's heartsick pining for him. And in the case of the "Italia" theme, the associations of F major and D minor fuse together to illustrate how Tom destroys his chances for happiness by further spinning out his web of lies.

On a more localized, micro-level of harmony, Tom's thwarted attempts at happiness are further represented by key moments in which the music is obstructed from resolving a dominant chord to tonic. Cadential frustration is used as a symbol of narrative frustration in this film (see Video 16 ✴ for a selected compilation of the following scenes). The first instance occurs when we are introduced to Tom's secret yearning for a higher station in life, in the concert performance of Beethoven's Op. 16 piano quartet (0:03:20). As Tom wistfully watches the performance, the dynamic level of the music swells as the camera zooms in on his face, to indicate how much the music means to him and how deeply he longs to be a legitimate audience member.[10] Just when the quartet plays a luxurious V^7 chord (✴), with tension swelling toward imminent resolution (the ♯2̂ chromatic passing tone in the first violin begging to resolve to 3̂), an irritated audience member turns around to glare haughtily at Tom, who instantly lets the velvet curtain fall shut

---

8. Referring to the Yared score excerpts on the supplementary website, notice that the melody-line pitches of the "Italia Corrupted" theme are virtually identical to those of the "Italia" theme, with the undergirding key being the only real difference (the starting pitch of A functions as 3̂ in F major and 5̂ in D minor). This clever repurposing of the melody gives the sense of a single person operating under very different circumstances.

9. Meredith is in love with Tom—whom she knows as "Dickie"—and Tom strings her along to cover his tracks.

10. The increased dynamic volume of the music is the audio equivalent of the camera zoom-in. It directs our attention to what the filmmaker wants to emphasize.

with a thud. That C♯ (♯$\hat{2}$) is forced instead to return back to C♮ across the scene cut, where it now functions as $\hat{5}$ of F major, as Tom plays the first movement of Bach's *Italian Concerto*. The frustration of the F dominant-seventh chord being forced to abandon its appellative ambitions and slink back to F major is analogous to Tom's frustration at being hampered from the life he longs to lead. Now playing the piano onstage after the audience has left, Tom's musical fulfillment is once again thwarted as a stagehand furiously throws on the house lights (with a violent cracking sound) to chastise Tom for touching the piano. Once again a dominant seventh chord (✶) is frustrated for Tom, as he is forced to leap away from the piano before he can resolve the dominant to tonic.

In a following scene, Charlie Parker's "Ko-Ko" (in B♭ major) finds Tom readying his meager wardrobe for the trip to Italy (0:06:09). Tom blurts out his disgust ("Ugh!") after the piece ends abruptly on a terse dominant (F), as yet another obstructed, resolutionless dominant grates on his nerves. This hanging dominant F finds resolution by proxy a few seconds later, when "Mache dich, mein Herze, rein" from Bach's *St. Matthew Passion* enters in the key of B♭ major, at the exact moment Dickie's photograph (in a Social Register book) is shown in extreme close-up.[11] This is symbolic of Tom trying to cobble together resolution/fulfillment in his life by mimicking (and eventually usurping) Dickie's identity.

Miles Davis's rendition of "Nature Boy" accompanies the bath scene, as Dickie soaks in the tub with Tom sitting (fully clothed) beside him (0:31:16). Palpable erotic tension is present (it's not yet entirely clear if it is one-sided or mutual), and Tom makes his first concrete attempt to sexualize their relationship. Tom's desire is conspicuous as he leers at Dickie's naked body, but Dickie saunters flirtatiously away leaving Tom frustrated once again. The G-minor music (✶) ends with ♭$\hat{2}$ in the bottom voice (bass) and $\hat{2}$–$\hat{3}$ motion in the upper voice (trumpet), resisting the gravitational pull to resolve to $\hat{1}$, making yet another frustrated ending indicative of Tom's obstructed desire. Similarly, in the scene where Tom dances as he dresses himself in Dickie's attire (in the reverse of a striptease), Dickie walks in and disgustedly shuts off the record (Bing Crosby's "May I?") on a fully diminished-seventh chord (vii°7/ii), leaving Tom flustered and disappointed in yet another instance of a frustrated dominant function.

After Tom has usurped Dickie's identity and life, we see him playing Bach's *Italian Concerto* once again, this time on the shiny new Alfonsi piano he has purchased for himself (as discussed before). This recalls the earlier scene when Tom was a lowly restroom attendant sneaking a few illicit moments on stage after the

---

11. The text of this Bach excerpt, too, eerily foretells Tom's yearning to make a fresh start by taking over Dickie's life: "Make yourself pure, my heart, I want to bury Jesus myself" *(Mache dich, mein Herze, rein, Ich will Jesum selbst begraben)*.

concert to play the Bach, and now he has his *own* concert grand in his lavish new apartment. But Tom is once again prevented from attaining the satisfaction of resolution when Freddie's insistent pounding on the front door arrests his playing in the midst of a cadential six-four chord. He has managed to claw his way up the socioeconomic ladder, but somehow he is now even *further* away from attaining his goal of tonic resolution (arrested as he is on the cadential six-four, prevented from even resolving it to the dominant-seventh chord).

When Tom has started a relationship with Peter, he plays Vivaldi's *Stabat Mater* (in F minor) at a dolorous tempo in Peter's apartment. While confiding his feelings of eternal frustration, Peter airily walks up to the piano and brashly plays the mocking, cheeky final cadence of "Knees Up, Mother Brown" (1:46:38). After a moment of silence, Tom smiles wistfully and says, "I keep wanting to do that." He goes on to define "that" as "flinging the door (to his heart) open," but we can construe another meaning in this line, referring to Tom's ever-unfulfilled desire to reach cadential resolution—which Peter attained so effortlessly, in the key of C major. Significantly, this bit of C major played by Peter is the *only* time C major is heard in the entire film. This "pure" key (no accidentals) is available only to the pure-hearted Peter—hence the use of C major as a singular key in this film.

During Peter's performance of the *Stabat Mater*, the angelic voice of the boy alto evokes a sense of innocence as Tom gazes up at Peter with love and tenderness (reciprocated by Peter's smile), and he contemplates the glorious possibility of regaining his *own* innocence, in a fresh start with his new love.[12] However, Tom is once again cut off (by a scene change) before musical fulfillment is achieved, leaving the music hanging on a dominant chord. The final moments of the film return to the theme from "Lullaby for Cain," which is, in essence, the story of Tom (the lyrics referring to the biblical story of Cain murdering Abel). After the "Lullaby" has run its course, the "Death" theme—which oscillates solely between tonic and dominant—ends the film on a dominant harmony, as the black screen mask visually blots Tom out with a barn-door wipe. This half-cadence ending leaves Tom's fate open-ended: Will he get away with this latest murder, or won't he?

Tom's ever-elusive quest for cadential resolution is symbolic of his inability to achieve the life and love he seeks. The endless string of obstructed cadences leaves Tom forever in the realm of dominant function—within sight of his goal but unable to attain it. Cadential frustration in this film is used as a musical analogue for Tom's frustrated desires. Figure 23 shows a tonal graph of the deployment of important key areas in this film, and Figure 24a captures the same information in

---

12. Minghella oversaw an original recording of the Vivaldi for this film. The quality of the boy's voice is unusually pure and high compared to other recordings. In fact, it is unusual to find a recording of this work featuring a boy alto rather than a contralto.

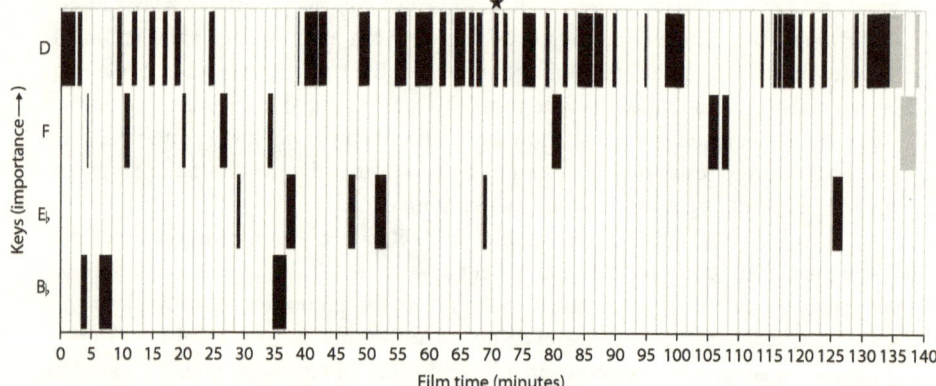

FIGURE 23. Tonal graph for *The Talented Mr. Ripley*. The gray bars represent cues during the closing credits (which is a framing element and not strictly part of the narrative). The star marks the germinal Tchaikovsky excerpt.

a different format in a tonal staff.[13] Parts b and c of Figure 24 reduce the "foreground" level (of part a) to reveal the symmetrical framework of the film's tonal design, which mirrors the symmetry of the narrative. The "middleground" level (Figure 24b) consists of the three most important keys, which form rotational cycles reflecting Tom's cyclical attempts at happiness: he begins in lonely darkness (D), strives for happiness (with Dickie) (F), meets with disappointment (E♭), and sinks back into loneliness (D). He strives once more for happiness (with Peter) (F), is again disappointed (E♭), and returns to darkness (D). Reducing the tonal staff even further yields the "background" level (Figure 24c), consisting of the tonal pairing of D minor and F major poles, reflecting the most basic binary of Tom's states: unhappiness and happiness. Even the closing credits emphasize the polar dichotomy of this tonal pairing by revisiting these two main tonics in a tonal coda, as shown (in gray) on the right-hand side of the staff.[14]

The tonal landscape of this film is built around the antipodal pairing of D minor and F major, and the significance of these keys originates in two preexisting works that are pivotal to the narrative. The key of D minor comes from Tchaikovsky's *Eugene Onegin*, significant for its tale of a man killing his most beloved friend. The key of F major comes from Bach's *Italian Concerto*, which symbolizes Tom's sanguine hopes for Italy. These two pieces were selected for the symbolic meaning

13. Mode is collapsed for the sake of visual clarity, as mentioned earlier. Refer to the "Logistics" section of chapter 1 for the full discussion of tonal graphs and tonal staffs.

14. Note that the most prevalent themes are absent from the closing credits, and a *new* cue is introduced—so the recurrence of these central keys in this tonal coda is not simply an artifact of recycling the main music of the film.

FIGURE 24. Three levels of tonal staff for *The Talented Mr. Ripley*. The gray noteheads represent cues during the closing credits.

a. "Foreground" level.
b. "Middleground" level.
c. "Background" level.

they impart to the narrative, and their keys serve as the foundation for the tonal design of the entire film.

THE GRAND BUDAPEST HOTEL (2014)

Techniques featured: associative tonality, transposition, intertextual tonality, meta key, tonal symmetry.

RELEVANT PLOT SYNOPSIS

In the present (unspecified) year, "Student" (Jella Niemann) enters the cemetery to pay homage to the monument of the great "Author" (Tom Wilkinson). As she looks down at his book, we flashback to "Author" (in 1985) narrating the prologue of his book. In the prologue he explains that this book will recount his experience at the Grand Budapest Hotel in 1968 (when he is known as "Young Writer").

Located in the Republic of Zubrowka (a fictional Central European state ravaged by war and poverty), "Young Writer" (Jude Law) checks into the Grand Budapest Hotel—a once-grand hotel that has fallen on hard times. "Young Writer" encounters the hotel's elderly owner, Zero Moustafa (F. Murray Abraham), who subsides into flashback (to 1932) to tell the tale of how he came to own this hotel, and why he is unwilling to shut it down.

Zero began as a lobby boy (Tony Revolori) in 1932, under the direction of the hotel concierge, Gustave (Ralph Fiennes). When one of Gustave's best clients, Madame D (Tilda Swinton), dies suddenly, Gustave takes Zero to attend the reading of her will. Kovacs (Jeff Goldblum), the executor of her will, reveals that Madame D has bequeathed a priceless painting to Gustave, which enrages her greedy children. With Zero's help, Gustave surreptitiously takes the painting, and he promises to make Zero his heir, in exchange for his help. Soon after, the eldest son Dmitri (Adrien Brody) accuses Gustave of murdering Madame D, and Gustave is arrested and imprisoned. Zero helps Gustave escape from prison by smuggling tools concealed inside cakes made by Agatha (Saoirse Ronan), the baker girl at Mendl's Bakery, and Zero's fiancée. Once Gustave escapes from prison, he and Zero are pursued by Jopling (Willem Dafoe), Dmitri's hired assassin. To facilitate their safe passage, Gustave calls upon the Society of the Crossed Keys, a fraternal order of concierges. Through the Society's facilitation, Gustave and Zero travel to a mountaintop monastery. Jopling follows them there, and a chase ensues. Zero pushes Jopling to his death and rescues Gustave.

Back at the Grand Budapest Hotel, war is looming, and the ZZ militia have commandeered the hotel as their barracks. Agatha and Zero bring a delivery of Mendl's pastries to the hotel, to try and retrieve the painting. Dmitri shows up at the same moment, and a chase and chaotic gunfight ensue. Zero and Agatha flee with the painting, which has concealed in its frame a copy of Madame D's will (which proves Gustave's innocence). The will also reveals that Madame D was the owner of the Grand Budapest Hotel, which she bequeaths (along with her fortune) to Gustave. Gustave appoints Zero as the new concierge, and Zero and Agatha marry.

After the war, Zubrowka is annexed. During a train journey across the border, Gustave is shot by ZZ soldiers when he attempts to defend Zero. Zero inherits the Grand Budapest Hotel from Gustave and vows to carry on his legacy. Some years later, Agatha dies of illness, and Zero is left alone, struggling to keep the dying hotel afloat as the country sinks further into economic depression.

Back in 1968, "Young Writer" is captivated by Zero's story and begins to formulate the book in his mind.

Back in 1985, "Author" completes his book.

Back in the present, "Student" finishes reading the book, in front of "Author's" monument.

The Grand Budapest Hotel is a sterling example of the poignant, nostalgic beauty evoked by Wes Anderson in his films. He is renowned for creating gorgeously detailed miniature worlds in which every aspect of the *mise-en-scène* is ornately stylized. Scholars and fans alike have long rhapsodized over the exquisite symmetry of Anderson's staging and cinematography, and this particular film extends his characteristic symmetry to the sonic level as well, in what turns out to be an ingenious reflection of the narrative.

Like two of the other Wes Anderson films we analyzed in chapter 3 (*The Royal Tenenbaums* and *Fantastic Mr. Fox*), this film also begins with the reverent visual introduction of an eponymous book cover (✳), paired with music and sound effects that invite us to climb through book cover and into the filmic diegesis. As the opening credits unfold, A-major music (we'll discuss this key soon) begins on a tonic harmony and progresses to a dominant-seventh harmony when the first visual scene of the film opens. To help smooth this transition from credits-space to narrative-space, the V⁷ chord of the nondiegetic music is corroborated by diegetic sound effects such as a church bell ringing on the chordal fifth of the V⁷ chord (0:00:54). This cue's final tonic resolution is synchronized with the camera's cut to a close-up of "Author" on the book jacket of his eponymous novel (titled *The Grand Budapest Hotel*), which takes us one stratum deeper into the layered narrative (into the year 1985, where "Author" will introduce his book to us).

This A-major synchronization of cadence and author is mirrored in the *closing* sequence of the film: a church bell once again tolls the chordal fifth of the music's V⁷ chord, joined this time by a car horn honking the chordal seventh (the car horn even matches the pitch and rhythm of the top vocal line) (1:33:31). The 2̂ of the church bell ends just before the music cadences to tonic, so 2̂ is heard to resolve smoothly to 1̂. This time, the tonic resolution is synchronized with presentation of the author Stefan Zweig (✱)—upon whose writings Anderson modeled this film. Thus tonic resolution in the *opening* sequence is synchronized with homage to the *diegetic* author, while tonic resolution in the *closing* sequence is synchronized with homage to the *nondiegetic* author (these opening and closing sequences are shown in Video 17 ✱).

The soundtrack for *The Grand Budapest Hotel* utilizes relatively few preexisting works (in comparison to Anderson's other films) and relies mainly on a large network of original themes composed by Alexandre Desplat. Unlike most films, which feature a small number of prominent keys amid a light static of unimportant[15] keys, *The Grand Budapest Hotel* makes very concentrated use of a small number of keys: F, G, A, B♭, and C (major and minor) are the only keys used throughout this entire film. First let's look at how each key is used in this film, and then we will examine how the keys relate to one another.

The structure of this film consists of three distinct time periods nested within one another: from the present it flashes back to 1985, then to 1968, then to 1932, and ends by traversing this temporal progression in reverse, so that the narrative levels fold/unfold symmetrically into one another like nesting dolls. As Table 14 demonstrates, the introduction of each of these three time periods is corroborated both visually and sonically. Visually, the three time periods are represented by three distinct screen aspect ratios in which those portions of the film are shot. Sonically, each of these time-period/aspect-ratio pairings is introduced by a distinct key: screen size 1.85 and year 1985 are bookended by the film's first and last instance of A-major music, the key of C major enters and leaves the soundtrack with screen

---

15. *The Talented Mr. Ripley*, for instance, contains cues representing all twelve tonics, but only D, E♭, F, and B♭ are of narratival and durational importance in the film. The "other" (narratively unimportant) keys in *The Talented Mr. Ripley* account for 19 percent of the film's total musical time (see Table 12), whereas the "other" keys in *The Grand Budapest Hotel* only add up to 0.4 percent.

TABLE 14 Introduction of the three time periods, screen aspect ratios, and keys in *The Grand Budapest Hotel*

Time Period	Screen Aspect Ratio	Key
1985	1.85 : 1	A major
1968	2.35 : 1	C major
1932	1.37 : 1	B♭ minor

TABLE 15 A-major cues (all traditional folk tunes) used in paratextual framing elements of *The Grand Budapest Hotel*

Location	Cue	Key	Type
opening credits	"S'Rothe-Zäuerli"	A major	preexisting
opening sequence	"S'Rothe-Zäuerli"	A major	preexisting
	(... main body of film ...)		
closing sequence	"S'Rothe-Zäuerli"	A major	preexisting
closing credits	"Kamarinskaya"	A major	preexisting
	"Svetit Mesyats"	A major	preexisting

size 2.35 and year 1968, and B♭ minor is paired with screen size 1.37 and year 1932.[16] Thus both screen aspect ratios and keys are used as structural parameters to delineate the nested stages of narrative in this film.

The first music we hear in the film is the lovely Tyrolean chorale "S'Rothe-Zäuerli" in A major (0:00:36). This key accompanies the opening credits and opening sequence of the film, during which we enter Author's book. The next time we hear music in A major is the *closing* sequence of the film, in which we emerge from Author's book and on to the closing credits (1:33:11). As Table 15 shows, the key of A major does not appear anywhere in the body of the film but is used exclusively for the paratextual framing elements. In both the opening and closing frameworks, the A-major tonality consists of nondiegetic music and diegetic sound effects intermingling together to blur the boundary of diegesis between credits-space and narrative-space.[17]

Like A major, the key of C major also plays a non-associative role in this film, serving as a device of narrative advancement. C-major music propels each of the

---

16. The screen aspect ratios and keys are paired when they (and the time periods) are *first introduced* and *last exit* the film, but the ratio-key pairing does not perpetuate through every subsequent scene (i.e., the screen aspect ratio does not shift to 1.37 every time the key of B♭ minor is used).

17. As Winters (2012) discusses, the blurring of diegetic boundaries is also a hallmark of Wes Anderson's *narrative* (i.e., not just musical) style.

TABLE 16 Introductions and alterations to character relationships in the key of
C major in *The Grand Budapest Hotel*

Relationships *Introduced*	C-major Music	Timestamp
Author and his audience	"The Alpine Sudetenwaltz"	0:02:58
Author and Young Writer	"Mr. Moustafa"	0:03:37
Young Writer and Mr. Moustafa	"Mr. Moustafa"	0:08:28
Mr. Moustafa and Zero	Vivaldi Concerto for Lute and Plucked Strings—I. Moderato	0:14:23
Gustave and Agatha	Vivaldi Concerto for Lute and Plucked Strings—I. Moderato	0:47:06

Relationships *Altered*	C-major Music	Timestamp
Zero and Gustave	"The War (Zero's Theme)"	1:01:15
Gustave and Zero+Agatha	"Third Class Carriage"	1:10:08
Zero and Gustave	"Cleared of All Charges"	1:26:55
Zero and Agatha	"The Mystical Union"	1:28:07
Young Writer and Mr. Moustafa	"Mr. Moustafa"	1:31:15

film's three distinct narratives by underscoring the *introduction* of core characters and the *alteration* of relationships between them (Table 16). C major *introduces* the relationship between the Author and his audience (when he makes direct eye contact and addresses us), between the Author and the Young Writer (when they are revealed to be the same person), between the Young Writer and Mr. Moustafa (when they meet), between Gustave and Agatha (when Zero brings her home to meet him). C major *alters* the relationship between Zero and Gustave (when Gustave first bonds with Zero), between Gustave and the young couple (when they ask Gustave to officiate their wedding), between Zero and Gustave again (when Gustave promotes Zero), between Zero and Agatha (when they get married), and finally between the Young Writer and Mr. Moustafa (when they part ways, after the Young Writer's understanding of Mr. Moustafa is complete).

The key of B♭ minor is used associatively to depict Madame D and all the turmoil spawned by her mysterious death and the ensuing danger-filled chase that propels the 1932 narrative—the *Sturm und Drang* "development" section of the film, metaphorically speaking. From the moment Madame D is introduced, clinging mawkishly to Gustave, to the news of her sudden suspicious death, the midnight train ride to her Lutz estate, the rapacious reading of her will, Gustave's absconding with her priceless painting, and her children's long and winding pursuit of Gustave—B♭ minor underscores every element of this storyline, in the form of numerous different themes by Desplat. Recall from chapter 3 that Wes Anderson's *Fantastic Mr. Fox* also featured B♭ minor as the villain key (associated with Boggis, Bunce, and Bean), so this feels like a nod of intertextual tonality.

The only other minor key in this film is C minor, which is used to portray despotic authority—namely Gustave's wrongful imprisonment and the occupation of the Nazi-like "ZZ" militia. This key is used when Gustave is pushing the mush cart around the prison at meal time (0:41:50), when fellow prisoner Ludwig describes his escape plan (0:43:45), when the prisoners make their escape (0:55:48), when Zero and Gustave run from the authorities (1:02:17), when the ZZ troops are stationed at the hotel (1:20:55), and when a shooting melee breaks out on the mezzanine amid a sea of ZZ banners (1:25:53).

G major is the sparkling key of Agatha in this film. The Vivaldi Lute Concerto is written in the key of C major, but internal G-major sections are used during Agatha-centric scenes and sequences. Sometimes the G-major passages are excerpted for Agatha, and sometimes the Vivaldi is used as a montage sequence, during which the music modulates to G major precisely when Agatha's scene in the montage begins. G major is first heard when Agatha is introduced in Mendl's Bakery (0:17:05): the Vivaldi has modulated to G major via ascending 5–6 sequence, the music is suspended while a (diegetic) ticking clock marks the 4/4 meter for two-and-a-half measures, and then the music starts back up in the G-major middle section of the movement. The second instance of G major begins directly at this middle section when Moustafa discusses his relationship with Agatha (0:46:00). The third instance occurs while Zero gazes up at Agatha on a pedestal (the carousel horse), professes his love, and proposes marriage (0:46:28). This scene features another G-major work, Strauss's "Roses from the South," playing diegetically as whimsical carousel music during a close-up of Agatha's face, portraying her ethereal, enchanting quality (✶).[18] The final occurrence of G major (the Vivaldi again) accompanies Agatha as she dauntlessly hides the tools for Gustave's prison escape inside of Mendl's pastries, and Moustafa emotionally lauds her bravery and devotion (0:47:50).

F major is the most important key in this film, and it is reserved exclusively for the resplendent majesty of the Society of the Crossed Keys. This key enters the soundtrack when Gustave begins contacting the other hotel concierges, mobilizing the Society in an elaborate chain of action (1:04:20). The "Society of the Crossed Keys" theme accompanies this montage sequence, and layered atop this theme is the song "Happy Birthday To You," sung in F major by a children's birthday party at M. Ivan's hotel (1:05:35). The final iteration of F major (the "M. Ivan" theme) occurs when M. Ivan picks up Gustave and Zero in his car (1:07:21).

F, G, A, B♭, and C account for all the musical cues (and even the pitched sound effects, discussed in chapter 5) in this film's soundtrack, and if we look at these five tonics together, we notice that they form the first five members of an F major meta

---

18. The original key of the Strauss composition is F major, so what we hear in this scene is a version *transposed* to the Agatha key of G major.

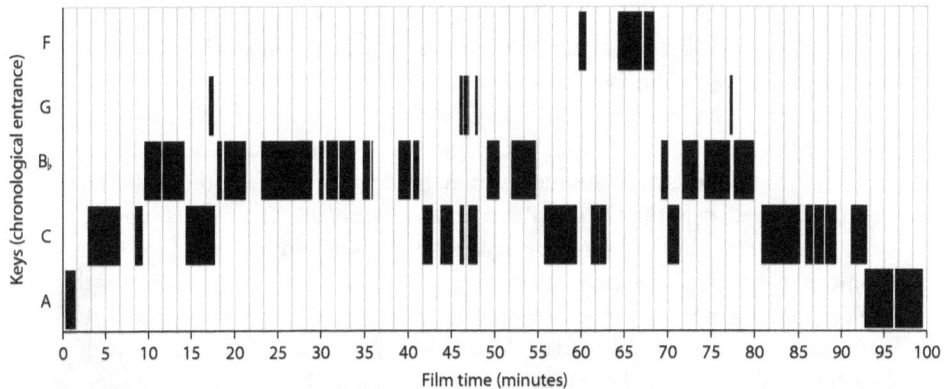

FIGURE 25. Tonal graph for *The Grand Budapest Hotel*. The keys are arranged according to their first chronological entry in the film.

key—which is the key of the Society of the Crossed Keys. The narrative germinates from the foundation of this proud Society, so F major can be seen as a sort of (conceptual) fundamental harmony in this film.[19] (Remember from chapter 3 that F major was the meta key in Anderson's *The Royal Tenenbaums* as well.) This meta key forms the framework of the film's tonal design, much as the stalwart Society serves as the framework upholding the hotel network. The tonal graph in Figure 25 illustrates the focal role of F major.

As you can see in this tonal graph, both the narrative and the soundtrack build up to the climactic apex of F major (during which the magic of the Society of the Crossed Keys is revealed), and then ramp back down after attaining it. And in a beautiful instance of tonal symmetry, the chronological order in which the keys first *enter* the film soundtrack is palindromic with the order in which they last *exit*:

<pre>
                F major
        G major   G major
      B♭ minor   B♭ minor
     C major   C major
    A major   A major
</pre>

---

19. The film romanticizes the autumn years of an antiquated, grand hotel, which only continues to function (along with its peers) because of the underlying fundamental network of proud, dedicated concierges. They are the beating heart of the gracious, old-world hotel system, and the revelation of the Society is the grand telos of the film.

This elegant formation creates a chiasmus, in which the key of the Society of the Crossed Keys stands at the center of its own cross of keys! And with this delicious tonal pun, the entire film forms a mirror image of itself.

The symmetry of *The Grand Budapest Hotel*'s tonal framework reflects the symmetry of the narrative: both trace out palindromes to begin and end in the same time and place. The apex of this tonal symmetry does not occur at the midpoint of the film but rather at the Golden Ratio point (Figure 26). The Golden Ratio has been revered since ancient times by mathematicians and mystics alike for its harmonious, aesthetically pleasing proportions. The structural arrangement of this ratio is manifested in nature at scales both large and small, from the nautilus shell to the shape of the Milky Way galaxy. The Golden Ratio has influenced the work of countless artists, such as Leonardo da Vinci, Salvador Dali, and Le Corbusier, as well as architecture from ancient through modern times. In film, people frequently find the proportions of the Golden Ratio in the visual dimensions of scenes crafted by auteur directors.

With regard to the *temporal* deployment of the Golden Ratio in film time (like what we observe in *The Grand Budapest Hotel*), Emily Verba (2012: 59) mentions two films by auteur directors in which structurally significant narrative events occur near the Golden Ratio point of the film's duration: the climactic Odessa Staircase scene (Act IV) in Eisenstein's *The Battleship Potemkin* (1925) and the second half (following the "Intermission" intertitle) of Kubrick's *2001: A Space Odyssey* (1968). The introduction of the Society of the Crossed Keys bears similar structural significance in *The Grand Budapest Hotel*, since this Society is the keystone of the entire narrative—and the Society's key (F major) is likewise the keystone of the soundtrack. Of course, the timing of the Golden Ratio point in this soundtrack is not precise to the decimal point, but then again, the Golden Ratio is used approximately in *most* constructions of art, architecture, and music. Scholars continue to debate hotly over the occurrence of the Golden Ratio in human-made work,[20] but whether architects or auteurs are evoking the Golden Ratio explicitly or responding unconsciously to its pleasing proportions, it remains there for us to speculate about and interpret.

The tonal symmetry of *The Grand Budapest Hotel* soundtrack is the sonic counterpart to the meticulously sculpted visual symmetry of Anderson's cinematography and *mise-en-scène*, which is one of the hallmarks of his filmmaking style.[21] His exquisitely elaborate sets, props, and characters infuse the magic of a whimsical

---

20. Mario Livio (2002) does an admirable job of distilling thousands of years of Golden Ratio fervor into a concise, engaging history. However, he does not approach the subject as an objective observer, taking his myth-busting hammer to even creator-confirmed uses of the Golden Ratio and entreating humanity to "give up [the Golden Ratio's] application as a fixed standard for aesthetics, either in the human form or as a touchstone for the fine arts" (200).

21. See Thompson and Bordwell (2014) for a good example of one of the many discussions of Anderson's symmetry.

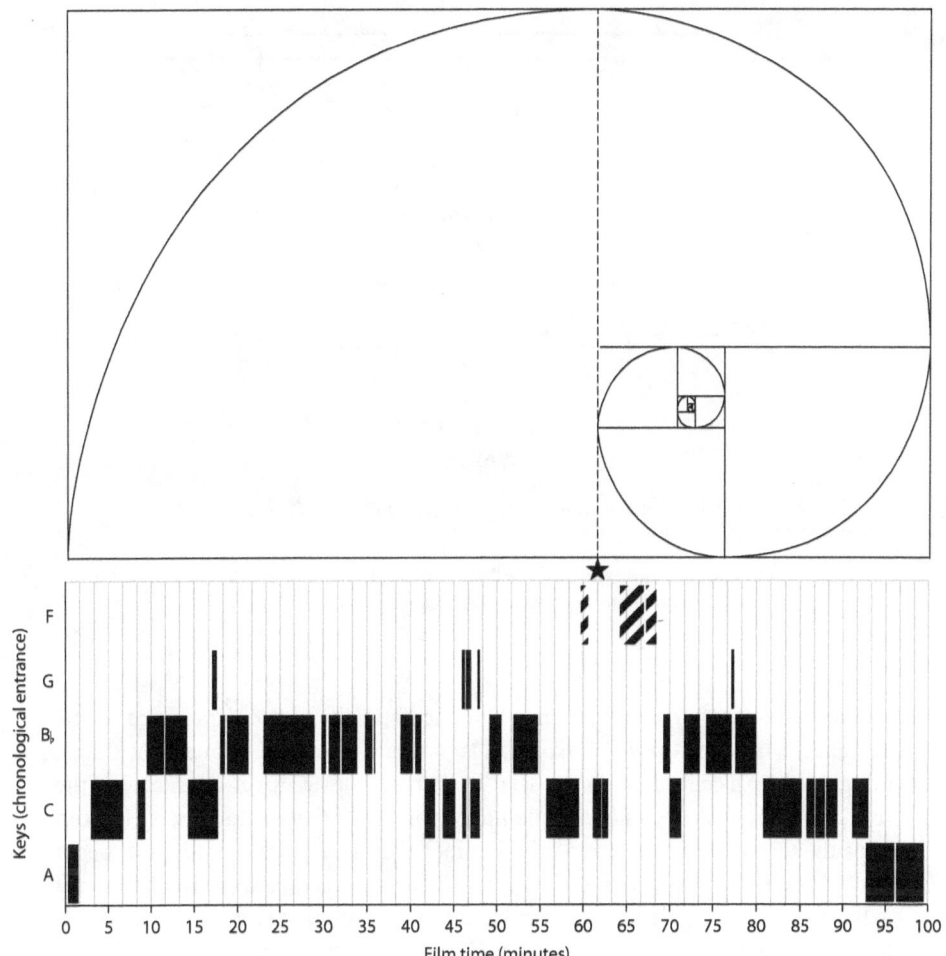

FIGURE 26. The fundamental key (striped) occurs at the Golden Ratio point (dashed) in *The Grand Budapest Hotel*.

fairytale kingdom into even the most quotidian settings. It comes as no surprise, then, that Anderson's musical *mise-en-scène* exhibits the same meticulous attention to detail and intricate filigree work.

COMPARISON OF THE TWO FILMS

On the surface these two films don't share much in common, but one major characteristic they do share is the flashback narrative structure. Although in *The*

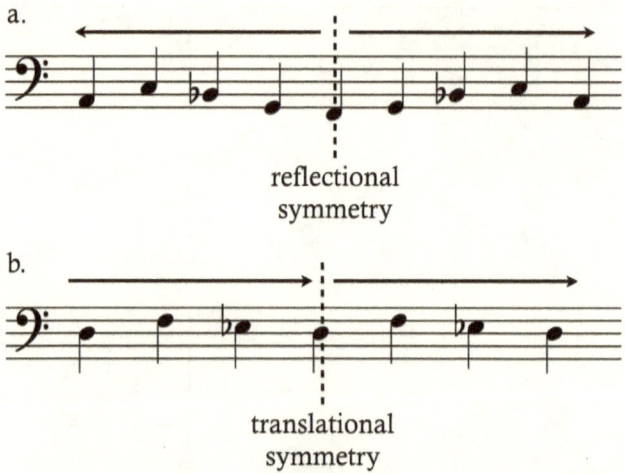

FIGURE 27. Tonal symmetry in *The Grand Budapest Hotel* and *The Talented Mr. Ripley*.

a. Reflectional symmetry in *The Grand Budapest Hotel*.
b. Translational symmetry in *The Talented Mr. Ripley*.

Talented Mr. Ripley one central narrator mediates the entire experience to the audience, while *The Grand Budapest Hotel* unfolds before the audience in stages, not guided by a central mediator. The "background-level" tonal symmetry of the soundtrack in these two films is the most striking similarity they share. In both films the tonal symmetry reflects the symmetry of the narrative: both plots begin and end in the same physical and temporal location, creating the sense that we are witnessing one expanded moment in time. But the nature of each symmetry is different (Figure 27). *The Grand Budapest Hotel*, with its palindromically telescoping layers, forms a *reflectional* symmetry. The fundamental key (F major) occurs at only one point in the film, highlighting the hallowed significance of the Society of the Crossed Keys (as well as the sense that the old-world hotel system peaks at this point, slowly declining from this point forward). *The Talented Mr. Ripley*, however, revisits the fundamental key (D minor) multiple times, following a pattern that forms a *translational* symmetry. This translational pattern (the "middleground" reduction from Figure 24b) reflects Tom's destructive pattern of making the same mistakes over and over again, by showing his inability to overcome the gravitational pull of his D-minor destiny.

In both films there is a central point within the narrative at which the tonal telos (the pivotal instance of the fundamental key) occurs. In *The Talented Mr. Ripley* this is the precise midpoint of the film (see Figure 23a) because the entire story has been carefully reconstructed (by Tom) after the fact. The scene from Tchaikovsky's *Eugene Onegin* imprinted powerfully on Tom, and it became psychological ground zero for his story. Because a passion for music is so central to his identity, Tom reframes his life story in terms of the opera—and that crucial D-minor event becomes the nucleus of Tom's tale. In *The Grand Budapest Hotel* the tonal telos occurs not at the film's midpoint but at the Golden Ratio point. Geometrically, the Golden Ratio represents a structure that subdivides infinitely into nested structures, and *The Grand Budapest Hotel*'s nested time periods, aspect ratios, and keys emulate that sense of structure-within-structure—almost as if zooming further in or out might continue to yield more nested narratives.

As we saw in these two films, the tonal design not only corroborates important narrative themes but also reveals filmic insights and subtleties we might have otherwise missed. For instance, knowing that the majority of Tom Ripley's music is in the key of D minor because of the *Eugene Onegin* excerpt at the film's midpoint shows us that this operatic moment imprints on Tom and recasts his memories through the lens of the opera's plot. Knowing that *The Grand Budapest Hotel*'s soundtrack is tonally symmetrical highlights the nested nature of the narrative, and the Golden Ratio point of its tonal zenith invites us to envisage infinitely more layers nested within. The remarkable use of tonal symmetry in *The Talented Mr. Ripley* and *The Grand Budapest Hotel* has enabled us to draw intriguing connections between two films that don't share much in common by conventional standards. And while large-scale tonal symmetry might be unremarkable in a piano sonata by Liszt or Beethoven, analyzing the use of common-practice techniques in the *un*common-practice setting of film broadens our conception of the soundtrack qua "composition" (while also broadening our notions of the techniques themselves).

CLOSING THOUGHTS

After having immersed ourselves in close analysis of film soundtracks over these past three chapters, allow me to revisit the questions I posed at the end of chapter 1. Does key matter in film, and do we gain anything from considering the tonal layout of a soundtrack? (For example, the symmetrical arrangement of keys in *The Grand Budapest Hotel*, with the key of the Crossed Keys at the center of the cross of keys.) Would we lose anything if a particular cue was displaced to a different key? (For example, if Tom's "tragic" themes were written in a different key than the critical *Eugene Onegin* excerpt.) Surveying the tonal treasures we have amassed during our analytical expeditions, I believe we have answered these questions. Having focused almost exclusively on music in chapters 2, 3, and 4, let us turn now to sound effects in chapter 5, in which we explore the intersection and interaction between musical and sound design, focusing on the important role pitched sound effects can play in shaping the tonality of a soundtrack.

5

Unheard Sound Effects

The title of this chapter is, of course, a play on the title of Claudia Gorbman's seminal book *Unheard Melodies* (1987), which made us realize how unconsciously we perceive music in film and taught us to pay deliberate attention to the work music does in film. The purpose of my chapter here is to do the same for pitched sound effects in film. As I demonstrate, sound effects are often carefully pitched to interact with the tonality of musical cues in a film—to quite an astonishing degree, actually—but we don't really *register* the pitches of sound effects or hear them as musical elements. Thus I propose a new way of viewing-hearing films in which we perceive sound effects as integrated tonal members of the musical soundscape.

Chapters 2–4 concentrated on music, but here in chapter 5 we focus on the contributions sound effects make to the soundtrack and the narrative. I have broadly categorized some ways in which sound effects can be pitched to interact with music: opening sounds, associative sound effects, stepwise sound effects, tonally functional sound effects, musicalized sound effects, transposed sound effects, and simply concordant sound effects. We revisit some films from previous chapters to explore the role sound effects played in those contexts and also look at examples from several new films. (Films not explored in previous chapters have the release year listed.) The chapter concludes with a full-length analysis of *Baby Driver* (2017), a film in which sound effects are central to both the soundtrack and the narrative.

OPENING SOUNDS

Sound effects can be highly effective in the opening sequence of a film, before music and visual imagery usurp our attention. As you will recall, both *The Royal*

Tenenbaums and *The Grand Budapest Hotel* begin with pitched sound effects that help draw the viewer into (and *out of*, in the latter) the narrative (see Video 9 ✻ and Video 17 ✻). In both films diegetic sound effects (emanating from elevators, car horns, and church bells) contribute to the nondiegetic music's V^7 chord, which resolves to tonic just as we officially cross the threshold of the filmic narrative. Sound effects from the diegetic realm amalgamate with music from the nondiegetic realm to create a tonality that merges together the two worlds by transcending the boundary between them.[1] And the dominant-tonic resolution is synchronized with entrance into the narrative, to draw the viewer-listener through the looking glass and into the world of the film.

As we saw in *The English Patient*, acousmatic nondiegetic sound effects are used to establish an inscrutable soundscape and aurally convey the enigma of the mysterious stranger. The layers of nondiegetic and diegetic music in this film's opening sequence imply a D-centered tonality, but the pitches of the sound effects interact with the music to veil the precise nature (mode) of that tonality. Later in this chapter, we explore an excellent example of an opening-sequence sound effect that *sets the tone* (literally) for the entire narrative in the 2017 film *Baby Driver*.

ASSOCIATIVE SOUND EFFECTS

Once you start paying attention to sound effects, you notice that they are often pitched to correlate with other tonal design elements in the soundtrack. Since associative tonality is one of the most common components of tonal design, you find an abundance of associatively pitched sound effects. They can occur during concurrent music in that key, or on their own (i.e., in the musical "silence" between cues). And in some cases a pitch *itself* can take on association without correlation to a musical key (i.e., F-pitched sound effects taking on associative meaning without the key of F major/minor having any associations in the *music*).

The English Patient is an example of the latter (sound effects leading the charge in creating associative meaning), in that the C♯ *pitch* is used to signal impending danger, with virtually no contribution from the C♯ *key*. Leading up to the explosion of Jan's jeep (Hana's best friend in the Red Cross unit), another jeep horn honks on C♯ (0:10:45), matching the C♯ sung by an old woman in the street at the same moment. Shrill C♯ honking and rumbling C♯ engine sounds build up in volume and become foregrounded just before the jeep explodes (0:11:41). In the following scene, metal detectors hum loudly on a low C♯ as Kip searches the road for additional landmines (0:12:29). The C♯ pitch's most dramatic usage occurs during the flashback story of how Caravaggio lost his thumbs. An air raid siren on C♯ blares

---

1. As mentioned in chapter 4, the blurring of diegetic boundaries is also a hallmark of Wes Anderson's *narrative* (i.e., not just musical) style.

out when Caravaggio belligerently removes his mittens in response to Almásy's questioning, which leads to the flashback of his torture interrogation (1:34:30).

This entire flashback scene is saturated in C♯ pitches. A phone in the interrogation room rings jarringly on C♯, beginning with the first escalation of tension when the interrogator accusingly spits out Caravaggio's moniker "Moose" (which incriminates him in the records they have collected). The pitch C♯ is joined by A♯ and E—produced by flies buzzing about the room—to form a tense, suspenseful diminished harmony as the questioning intensifies. The sound of a passing plane engine humming on C♯ enters when Caravaggio realizes how hopeless his case is and makes a desperate attempt to appeal to their mercy.[2] The phone rings again on C♯ when the interrogator begins talking about cutting off Caravaggio's fingers, and a hollow, thumping, echoing sound on C♯ swells in volume after all other sounds have died out, as the interrogator pointedly asks if thumbs are fingers (leading to the grisly severing of Caravaggio's thumbs). There is *one* instance of danger-laden C♯-major music (Benny Goodman's "One O'clock Jump" plays while Hana dances with Caravaggio (2:00:21), anempathetically building up the tension just before a mine explodes and kills Sergeant Hardy), but otherwise the C♯ association with danger is entirely built by sound effects.

Also in *The English Patient*, a plethora of B-pitched sound effects during the sandstorm scene evoke the mysterious associations of the B-minor key in this film. Recall that B minor in this film is associated with the Count Almásy/"English Patient" character and the enigma of his amnesic identity. When a sandstorm begins brewing during the group's trek through the Sahara desert, the storm is initiated with whistling wind and metallic clinking sounds, all B pitches (0:58:42). The sandstorm traps Katharine and Almásy together in a jeep overnight, and all of the B-pitched sound effects cease the moment they begin bonding (with music in G major, the romance key). After they (and the storm) are done, B takes over once again, in the form of the "Oriental Theme" in B minor, and their window of intimacy closes back up. Thus this brief G-major intimacy is engulfed by the swirling sonority of B—preceded by B-pitched sound effects and followed by B-minor music—a miniature portrait of the *Leittonwechsel* relationship that tonally conveys Katharine and Almásy's tragic relationship.

In *The Grand Budapest Hotel* numerous sound effects are pitched to contribute to the villain key of B♭ minor. When Gustave describes Madame D's odious children and says they will dance on her grave, a train whistle blows on B♭ (0:19:14), which is î of the "Daylight Express to Lutz" theme. During this same theme, Police Inspector Henckels blows his whistle on B♭-D♭ (î-3̂) after Gustave demands that the officials release Zero (0:21:17). When Zero says he never knew his family, the close-up of Gustave's face is accompanied by a train bell on D♭ (0:34:13), which is 3̂ of the "Night

---

2. The plane engine begins on C♯, then descends as it flies past, due to the Doppler effect.

Train to Nebelsbad" theme. When Zero dashes away from Agatha's attic room, a church bell tolls on C (0:51:47), followed immediately by a grandfather clock chiming on B♭ (as Kovacs comes down the stairs), which together serve as $\hat{2}$-$\hat{1}$ of the immediately ensuing "The Cold-Blooded Murder of Deputy Vilmos Kovacs" theme. Just as this theme ends and Dmitri emerges from the Kunstmuseum (after murdering Kovacs), a timpani roll on F and the D♭-E♭ of a church bell coincide to function as $\hat{3}$-$\hat{4}$-$\hat{5}$ (0:54:52). When Gustave and Zero take the train to Gabelmeister's Peak, the train whistles on B♭ (1:13:00), which is $\hat{1}$ of the "The Cold-Blooded Murder of Deputy Vilmos Kovacs" theme. When Gustave and Zero return to the Grand Budapest Hotel, the motorcycle engine hums on B♭ (1:20:40), which is $\hat{1}$ of the previous "Canto at Gabelmeister's Peak" theme. And so on and so forth. The interaction of all these B♭-centric sound effects with the B♭-minor music helps convey the pervasive power and reach of the many villains in this film, as they relentlessly pursue the protagonists at every turn. That even ambient sounds align with the villains' sinister music makes them feel omnipotent and omnipresent.

On the opposite side of the spectrum in *The Grand Budapest Hotel*, the heroic key of F major is likewise fortified by F-pitched sound effects. During the climactic "Society of the Crossed Keys" montage (1:04:20), each Society concierge is accompanied by a different F-centric sound effect, layered prominently atop the concurrent F-major music (Table 17): Concierge George flashes his phone switchhook on $\hat{3}$,[3] Concierge Dino's phone rings on $\hat{3}$, a fire alarm rings on $\hat{1}$-$\hat{3}$ while Concierge Dino oversees the firemen, and Concierge Robin's phone rings on $\hat{1}$-$\hat{3}$.[4] There are *two* F-major cues during this montage, and they are listed in the same box because they occur concurrently: Concierge George conducts a group of children diegetically singing "Happy Birthday," and this is layered atop the nondiegetic "Society of the Crossed Keys" music. In this scenario the tonal alignment of diegetic sound effects, nondiegetic music, and diegetic music conveys the meticulous order and organization of the Society of the Crossed Keys, an elite network whose professional success is contingent on impeccable coordination and control—right down to the harmonization of their fire alarms and children's birthday parties!

Similarly, *Fantastic Mr. Fox* also features a climactic montage in which pitched sound effects are tonally coordinated to maximize the dramatic effect. The fox-versus-farmer showdown at high noon in the town square is peppered with sound effects that are harmonically and metrically coordinated with the concurrent music in D minor, the nemesis key (Table 18). As close-up shots of wristwatches show the

---

3. On early candlestick telephones, the switchhook was the forked cradle in which the receiver rested. When the user lifted the receiver from the switchhook, the telephone line became active. The user "flashed the switchhook" by tapping it a few times (making a dinging bell sound), to get the telephone operator's attention.

4. In the last leg of this chain of phone calls, the concurrent music briefly lands on a D-minor chord, and Concierge Martin's phone rings on $\hat{1}$ (D) to match this chord.

TABLE 17 Pitched sound effects and their roles in contributing to the concurrent F-major music in *The Grand Budapest Hotel*

Sound Effect	Pitch	Role	Plot Event	Concurrent F-major Music
phone switchhook	A	$\hat{3}$	Concierge George flashes the switchhook	"The Society of the Crossed Keys"
phone ring	A	$\hat{3}$	Concierge George calls Concierge Dino	
fire alarm	F-A	$\hat{1}$-$\hat{3}$	Concierge Dino oversees the firemen	"Happy Birthday"
phone ring	F-A	$\hat{1}$-$\hat{3}$	Concierge Dino calls Concierge Robin	

TABLE 18 Pitched sound effects and their roles in contributing to the concurrent D-minor/major music in the final showdown in *Fantastic Mr. Fox*

Sound Effect	Pitch	Role	Plot Event	Concurrent D-minor/major Music
watch alarm	D	$\hat{1}$	clock strikes 10:00	"Great Harrowsford Square"
watch alarm	A	$\hat{5}$	clock strikes 10:00	(in D minor)
laundromat dryers	A	$\hat{5}$	townspeople await the showdown	
fire truck alarm	D	$\hat{1}$	mayhem in the town as fires break out	
helicopter	D	$\hat{1}$	helicopter follows Mr. Fox's motorcycle	Georges Delerue "Le Grand Choral"
helicopter	A	$\hat{5}$	helicopter follows Mr. Fox's motorcycle	(in D major)

animals noting the precise start-time of the showdown, one watch alarm beeps on $\hat{1}$ (during a i chord), followed by another watch alarm beeping on $\hat{5}$ (during a V chord). The camera pans across townspeople awaiting the showdown, and clothes dryers in a laundromat hum around $\hat{5}$ (during a V chord). When fires and mayhem break out around the town, the fire truck alarm rings out on $\hat{1}$. And as Mr. Fox heads out on his motorcycle (to rescue Kristofferson), a "Bean Air" helicopter's rotor whirs first on $\hat{1}$ (during a I chord), then on $\hat{5}$ (during a V chord). All these D-minor-centric sound effects highlight the danger posed by Mr. Fox's nemeses, and they are sonically foregrounded, which builds up the intensity of this climactic sequence.

Also in *Fantastic Mr. Fox*, the B♭-minor key belonging to Boggis, Bunce, and Bean is frequently supported by B♭-pitched sound effects. A typewriter bell pings portentously on B♭ when Badger advises Mr. Fox not to buy the Boggis, Bunce, and Bean–adjacent tree house (0:08:07). This B♭ bridges a gap between the B♭-

minor music that precedes and follows it. When Mr. Fox dismisses Badger's stern advice, the same typewriter pings out a shrill B♭-B♮ bell (0:08:12), with the dissonance adding an exclamation point to the ominous warning. This B♭ omen comes to fruition when the B♭ of the Boggis Farms security floodlights, alarm bell, and alarm buzzer all go off to reveal Mr. Fox's burglary (0:19:41). The following morning, B♭ is used as a reference to this burglary: Mr. Fox's midnight chicken heist was accompanied by B♭-major music, and when Mrs. Fox questions him about the mysterious materialization of the chicken in their pantry, he evasively whistles an aimless tune (more like a whistled version of mumbling) in B♭ major (0:20:13), alluding to the B♭-major music that underscored his burgling of that very chicken. He tonally alludes to the burglary again when he whistle-mumbles in B♭ major after fibbing to Mrs. Fox about the chicken's "escape" from Boggis Farms (0:20:24).

In *Moonrise Kingdom*, as you will recall, Sam's key of G minor is commenced and concluded with G-pitched sound effects. The key of G minor is introduced by a G-pitched phone ring (0:11:24) as the camera zooms in for a close-up of Sam's picture (with the corresponding tonal wink shown in Figure 4). The key of G minor is bid adieu with a G-pitched bell toll (1:22:38) and makes way for the key of G major, the rainbow at the end of Sam's storm.

Our tonal analysis of *The Graduate* painted bright prospects for Benjamin, with his directional uprising from E♭ to E minor. The future outlook is promising for Elaine as well, and a crucial sound effect presages this prospect: when Elaine is facing her crossroad at the zoo (deciding whether to choose Smith or Benjamin), a train horn blares an E-B♭ tritone (1:20:15). This tritone (the only exposed tritone in the film) signals that Elaine must choose which path to take—her mother's or the unknown—and the E and B♭ of this tritone collision carry specific meaning in the film. We already know that the key of E minor represents a disruption of Benjamin's path from his starting key of E♭ minor—and as it turns out, the key of B♭ major represents a disruption of *Elaine's* path. The C-major song "Mrs. Robinson" (her mother's name) looms over Elaine like an oppressive forecast, taunting her to follow in her mother's footsteps. The key of C major is first introduced by a clock tower that interjects insistently in C major every time Elaine and Benjamin discuss marriage (1:28:21), followed by the C-major "Mrs. Robinson" when Elaine tells him she *might* marry him (1:30:27) and again when she tells him she won't (1:34:16). But significantly, the key of "Mrs. Robinson" is transposed from C major to B♭ major when Benjamin goes into full throttle (first driving then sprinting) to stop Elaine's wedding (1:36:39). C major saw Elaine on the conventional path to early marriage and her mother's despondent fate, and B♭ major sets the disruption of her wedding—and her fate—in motion.

The key of C major lodges one more feeble attempt to restore convention with a claustrophobic and ineffectual rendition of Mendelssohn's "Wedding March" (played by a comically nearsighted and ham-fisted church organist ✶) (1:42:22), which brays banally in C major until Benjamin successfully wrecks the wedding.

The significant exchange between Elaine and her mother during the turmoil of fleeing the altar (Mrs. Robinson: "It's too late." Elaine: "Not for me!") implies that Elaine has broken free of her mother's fate. She does *not* marry the mundane "right guy" (befittingly generic with the quintessentially nondescript surname "Smith"), drop out of college, and surrender her dreams, ambitions, and self—as her mother did.[5]

Elaine's bold choice, coupled with Benjamin's new key, insinuates that they may have successfully escaped the pull of the Pasadena country club set. Elaine will go back to Berkeley and finish college, Benjamin will likely join her there and maybe get a master's degree in something he actually cares about (rather than going into "plastics" as his elders decreed), and even if/after they break up, they will have aided one another with a *gravity assist* to escape the trajectories they were stuck in before.[6] And that E-B♭ sound effect was key in foretokening this mutual assist. That E-B♭ train horn signals the derailment of Benjamin's and Elaine's trajectories through their mutual entwinement: she helps him escape his father's life (E♭→E), and he helps her escape her mother's life (C→B♭).

STEPWISE SOUND EFFECTS

As with associative tonality, stepwise tonal relationships can very easily be reinforced by stepwise sound effects. There are any number of naturally occurring sound effects featuring a major- or minor-second relationship, such as phone rings, doorbells, intercom buzzers, and alarm clocks, not to mention sound effects that can be customized by sound designers.

The main tonal design element in *Breaking and Entering*, you will recall, is the C minor/D minor stepwise conflict signaling Will's conflicted feelings between his wife and his mistress. Will's dilemma is corroborated by an important doorbell that stridently buzzes the pitches C-D (0:54:47). Significantly, this is mistress Amira's doorbell—which *only* Will rings—so each time Will comes to his mistress's house, his guilty conscience is needled by this whole-tone tension. At the midpoint of the film the crucial montage sequence discussed in chapter 2 (during which both of Will's women are shown separately discussing his happiness) pulls out a Picardy-third resolution to end the C-minor music on a happy C-major triad. But *immediately* upon reaching this serene C-major chord, the jarring doorbell buzzes harshly on C-D (1:06:09), which once again takes Will back into his conflicted state of tension (see Video 18 ✷).

Likewise in *The Darjeeling Limited*, the trope of the stepwise clash is represented in a plethora of half- and whole-step-related sound effects. These sound effects are

---

5. Well, technically the marriage vows *were* completed, but the bride's exiting the church with another man does imply that an annulment is forthcoming.

6. A *gravity assist* is a maneuver in which one astronomical body alters its course by harnessing the gravitational pull of another to slingshot past.

strategically employed to signal moments of liminality between conflicting states. The entire troubled courtship between Jack and Rita, for instance, is aurally littered with step-related sound effects. The dining-car order bell juxtaposes a D♯ against the C♯-major music when Jack first notices Rita (which draws him apart from his brothers and draws her apart from her boyfriend) (0:11:02). When they first lock eyes and make a visual pact, the train horn is foregrounded with a blaring E♭-F dissonance (0:15:39). As their illicit relationship reaches its (strained) climax, a train horn sounds on G♯-A when Jack tells Rita he wants to kiss her (0:38:38), a buzzer sounds on A♭-B♭ when Rita says she plans to break up with the Chief Steward (0:38:58), the train horn blows on E♭-F when Jack tells Rita she *might* be important to him (0:39:06), and the buzzer buzzes again on A♭-B♭ as Rita quickly ejects Jack when the Chief Steward comes looking for her (0:39:14). As it turns out, the Chief Steward was the person ringing that angry A♭-B♭ buzzer, so he exerts his agency in this love triangle by provoking that petulant major-second in order to separate Jack and Rita (since Rita must always instantly come running when she hears the buzzer calling her to service).

Another scene fraught with seconds is the brothers' visit to the temple (0:23:17), which occurs at the height of their disunity. A myriad temple bells tinkle on C♯, while the temple devotees chant on D♯, and each brother rings the large entrance bell on D♯ as they cross the temple threshold. As they sit down to pray, a piercing tapestry of C♯- and D♯-ringing bells highlights the brothers' discordance as they clash with one another (in this place designed for peace). Later in the film, when the brothers are attending the village boy's funeral, bells ring on C♯ and D at the precise moment they transition from their flashback funeral (of their father) to their present funeral (of the village boy) (1:02:18), signaling their liminal emotional presence at both funerals at once (Figure 28). And to further emphasize the half-step nature of this moment, the twenty seconds leading up to the scene transition between past-funeral and present-funeral is filled with half-step alternation between the pitches E-F (and B♭-A) in the fourth movement of Beethoven's Symphony No. 7 (as seen in the score reduction in Figure 28) (see Video 19 ✷).

The trope of the dissonant stepwise relation in *The Darjeeling Limited* roguishly recurs as the end of the closing credits: the very last sound we hear in this film is the blaring E♭-F of the train horn (1:31:10).[7] This piquant dissonant-second ending puts a pragmatic twist on the conventional "happy ending," in which all problems are resolved and the characters go forth in perfect harmony. While these three brothers have reached a new level of harmony and understanding, their harmony will *undoubtedly* continue to be peppered with the dissonant clashes inherent to their siblinghood, as this final stepwise flourish reminds us.

---

7. This puts me in mind of Wagner's *Die Walküre*, which begins in D minor and ends in E major, and which replicates this large-scale harmonic relationship on a *localized* level with its final harmonic progression: a D minor triad moving to an E major triad.

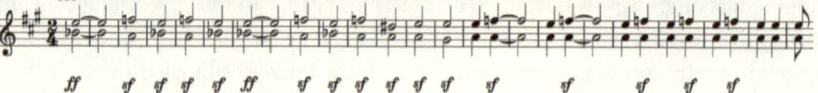

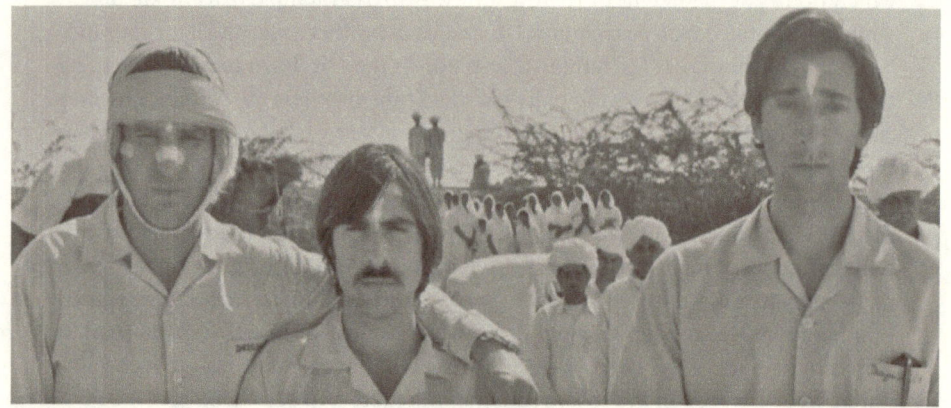

FIGURE 28. Transition from the flashback of their father's funeral (top) to their presence at the village boy's funeral (bottom). The father's funeral is accompanied by the alternating E-F pitches (top voice) of Beethoven's Symphony No. 7 (reduction of fourth movement, mm. 330–352), and the boy's funeral is accompanied by bells ringing C♯-D pitches. (Note how the mirrored visuals express their simultaneous presence at both funerals.)

TONALLY FUNCTIONAL SOUND EFFECTS

In the same way that the key of one cue can be used as dominant preparation for another key, sound effects have the power to portray harmonic function. Often dominant-functioning sound effects and cues are intermingled together, as the following examples show.

Functional tonality plays an important role in *The Graduate*, with abundant sound effects and cues that serve as the tonic or dominant of following or preceding cues. From the very opening of the film, Benjamin's airplane hums on a B♭ pitch (0:00:24), which serves as 5̂ of the ensuing cue ("The Sound of Silence") in E♭ minor (see Video 20 ✱). In the Taft Hotel with Mrs. Robinson for the first time, a C-minor cue (Dave Grusin's "Hotel") is soon after followed by a C-pitched hotel desk bell (1̂) (0:31:03) and a G-pitched telephone ring (5̂) (0:31:34). Their ensuing affair is accompanied by "The Sound of Silence" in E♭ minor (0:38:15), which dovetails with and serves as the dominant of Simon & Garfunkel's A♭-major "April Come She Will" (0:41:18), as Benjamin grows listless and dissatisfied with the affair. Toward the end of the film, a brief G-major instrumental number (untitled) by Simon & Garfunkel (1:41:56) acts as dominant preparation for the risible C-major "Wedding March" that directly follows. And incidentally, the first time we hear "Mrs. Robinson" in this film, it is whistled inchoately by Benjamin just after dropping off Elaine from their first date, *in G major* (1:08:00), which provides anticipatory dominant preparation for the C-major version we hear later on when the song officially begins. This phenomenon of interconnected tonics and dominants highlights the sense of cause-and-effect that intertwines the characters in this narrative, with every action setting off a chain reaction that affects the fates of the others.

Woody Harrelson's quirky (and semi-autobiographical) *Lost in London* (2017) features an equally quirky example of a tonally functional sound effect during the jail scene. Woody (playing himself) lies miserably on the floor of his jail cell, beseeching God for "another chance" at redemption. Just then, Willie Nelson's "London" in E major (with its twangy, bluesy Mixolydian intro) begins playing nondiegetically (1:25:25), its intro vamping between a I chord and a bluesy minor v chord (see Video 21 ✱). The music is immediately joined by the sound of a clock in Woody's jail cell diegetically chiming midnight on a D pitch (patience, Willie fans—I will address your protest shortly). The clock's D pitch serves as the chordal seventh of the music's I chord, turning it into a V⁷/IV which resolves to IV. The nondiegetic song fades out on a IV chord and is replaced by a diegetic version of the song as the camera pans slowly upward to reveal Willie Nelson himself standing behind Woody in his jail cell. Willie, portrayed as a wise and fatherly God-type figure (and introduced by a IV–I plagal motion, with its sacred connotations), answers Woody's prayer and counsels him on how to achieve the redemption he seeks.

Of course, this clock chime is actually part of Willie's (1972) "London" recording, meant to evoke the midnight chiming of Big Ben, one of London's iconic city sounds (as confirmed by Willie's opening lyrics: "The streets are dark and quiet at London after midnight—Listen"). But the camera in this scene dips deliberately upward to show the jail cell clock (which reads midnight), implying that the sound is emanating diegetically from the clock in Woody's jail cell. Now, if the song featuring a clock chime was a hit Beatles song, we might not be able to make a strong case for the illusion of diegetic sound in this film scene, because the song would be *so* ubiquitously well-known that virtually any Westerner watching the film would instantly recognize it as part of the song. But wildly popular as Willie Nelson is, this particular song is not among his better-known works; its host album (*The Words Don't Fit the Picture*) was one of only three of Willie's seventeen 1970s albums that did not chart on *Billboard*'s "Top Country Albums" chart.[8] And if YouTube can be used as a metric for song popularity, "London," posted from the official Willie Nelson YouTube Channel, has only six thousand views as of August 2022, and it is the *one and only* posting of "London" on YouTube (as of this date). Thus the *Lost in London* film made a safe bet in assuming that viewer-listeners would be able to hear this clock chime as diegetic.

Even if one logically realizes that the sort of clock featured in Woody's jail cell is *not* the sort that would chime, this is just one of a myriad ways in which film viewers cooperate via unspoken compact of suspended disbelief to accept what filmmakers present to us as "reality." In another interesting distortion of reality, the *actual* pitch of Big Ben's hour chime is an E, which would have easily and naturally fit Willie's E-major song as a concordant 1̂. Willie's song features a D-pitched clock chime (rather than E) in the intro to plant it as a "promissory note" for the upcoming IV chord in the verse.[9] That chordal seventh clangs insistently throughout the entire thirty-three-second intro, its appellative power magnetically pulling the music (and the listener) to its resolution, just as it pulls Woody to *his* resolution and redemption.

The Favourite (2018), a luxuriously dark period film directed by Yorgos Lanthimos, features another wonderful example of the blurred line between sound effects and music in creating functional tonality. This film captures the riveting, tension-filled triangle between Queen Anne (Olivia Colman) and her two companions, Abigail Masham (Emma Stone) and Sarah Churchill (Rachel Weisz), vying relentlessly to win the Queen's favor. In the beginning, Sarah is firmly established as the Queen's most intimate and influential companion; but Abigail cunningly works her way into the Queen's good graces from her initial position as a scullery maid,

---

8. "Willie Nelson albums discography," https://en.wikipedia.org/wiki/Willie_Nelson_albums_discography#1970s.

9. The concept of the musical "promissory note" was first coined by Edward Cone (1982).

eventually supplanting Sarah and procuring her banishment. Abigail's path to power is paved with mysterious G pitches that occur somewhere between the diegetic and nondiegetic realms—sometimes as ostensibly ambient sound effects (though the source and nature of the sounds is never clear) and sometimes as foregrounded musical sounds that defy classification as music or sound effect.

The G pitches are first introduced as arrhythmic string pizzicato, padding quietly in the background when Abigail first enters the palace (0:03:45). They recur a few scenes later when Abigail begins her kitchen work, now joined by a variety of G-pitched sound effects of clanking of kitchen equipment. The nature of the G pitches intensifies significantly when Abigail officially begins her climb up the ladder by sustaining an injury that enables her to forge a relationship with the Queen (0:10:00): Abigail burns her hand with lye and seeks out healing herbs in the forest, which she then guilefully proffers to the Queen, to soothe her burning gout wounds. Abigail's screams are sound-matched with Queen Anne's across the scene change, sonically illustrating the connection between them. The intensification of the G pitches accompanying this sequence is striking: the pitches (produced by a bowed viola) are now clearly foregrounded and form a rhythmically precise pulse that feels acutely calculated. Across the aforementioned scene change, when the Queen is shown screaming, the bowed Gs are joined by robustly plucked Gs (occurring on beats 1 and 3 of the implied 4/4 meter, respectively), forming a sort of point-counterpoint between Abigail and Queen Anne.

The plucked Gs are replaced by a piano's (slightly firmer) Gs when Abigail purloins a palace horse to ride off to the forest in search of the herbs. As this procession of Gs continues for several minutes, the viewer-listener slowly starts to wonder if this might be *music* rather than just sound (though it feels a bit vague to pinpoint, because it is technically just a pulse of single pitch, which we don't usually identify as "music"). It turns out to be a composition by Luc Ferrari entitled *Didascalies* (✳), brilliantly employed to convey Abigail's covert march toward power. (The title of Ferrari's piece is the Italian term *didascalies*, which roughly translates to "stage directions"—eminently appropriate for Abigail's systematic manipulation of the Queen.) This is no longer a murky aural haze of hushed G sounds (as in the earlier scenes) but a steady and determined procession toward a purposeful but as-yet unidentified goal (at this point we haven't yet been acquainted with Abigail's ambitions, and we think she is simply an earnestly helpful servant). Abigail's herbal offering earns her a prodigious promotion to serve as Sarah's maid, which grants her access to the Queen (and other power players in the palace).

The resolute Gs of *Didascalies* return again just before the midpoint of the film (0:56:59), when Abigail surreptitiously enters the Queen's bedroom and ensnares her sexually, climbing into her bed to continue her climb up the ladder of power. The volume increases as we cut away from the bedchamber to the Prime Minister informing Sarah that the battle is about to begin; he is referring to the battle with

France, but Sarah senses another battle. The volume continues to rise as Sarah hurries toward the Queen's bedchamber, and just as she unlocks the door, the G pitch of *Didascalies* leads directly into the C-minor second movement of Schumann's Piano Quintet (✳) (1:02:14) (see Video 22 ✳). The Quintet swells as Sarah swings the door open to reveal a naked Abigail lying in the Queen's sleeping arms. This climactic moment is underscored by the key of C minor—which had not been used hitherto in this film—and the mysterious procession of G pitches leading up to it are now understood as an hour's worth of dominant preparation. They started out timidly and quietly, growing bolder and more determined, until finally discharging on this moment of C-minor denouement.

Shortly after this major coup, Abigail suffers a setback when Sarah dismisses her from her position, and the indistinct, unidentified G-pitched sound effects accompany Abigail once again as she schemes to find a way to remain near the Queen (1:04:06). The C-minor Schumann recurs when Queen Anne surprises Sarah by informing her that she has appointed Abigail as her new Maid of the Bedchamber, handing Sarah another decisive C-minor blow. In yet another example of dominant preparation, a sustained instrumental G pitch swells unswervingly after Abigail poisons Sarah's tea (1:11:56), leading directly into another C-minor work—Bach's Fantasia in C Minor, BWV 562—as Sarah leaves the palace to meet her further undoing (her collapse in the forest and subsequent kidnap by miscreants). After this, Abigail turns the Queen against Sarah and brings about her banishment from the land.

The key of C minor in this film is associated with Abigail's triumph over Sarah, and the G pitches pave the path to C minor. These mysterious G pitches slink in unnoticed at first—as did Abigail herself—blending seamlessly into the ambient bustle of the old castle. But the Gs subtly begin to organize into a systematic march toward a goal, gaining in volume and strength before finally discharging their full dominant-functioning power into decisive C-minor conquests. The subtle origin of these G pitches as innocuous, ambient sound effects (before blending into Ferarri's *Didascalies*) perfectly mirrors Abigail's stealthy machinations, with her feigned wide-eyed innocence (as wide as only Emma Stone's eyes can be) as she steadily plots her way to power.

And before we leave this category, let us not forget Mozart's masterful $\hat{5}$-pitched fart in *Amadeus*, which replaces the music's expected $\hat{1}$ in order to obstruct a PAC and cadentially humiliate Salieri.

MUSICALIZED SOUND EFFECTS

Sometimes music is momentarily backgrounded so that foregrounded sound effects can function as part of the music, harmonically and rhythmically coordinated so they act as musical instruments. This fluid category can overlap with the others, as demonstrated by *Lost in London* and *The Favourite*, both of which were included in the previous category but could just as easily have been included in this category; the main characteristic is that the sound effects feel integrated into the music—almost functioning as musical instruments—rather than a layer superimposed atop it.

Fantastic Mr. Fox features several examples that combine this category with the previous category, with sound effects that act as musical instruments and portray dominant function. For instance, the nerve-racking moment Mrs. Bean enters the cider cellar and the animals risk getting caught, the C-minor music cuts off on a V chord, and the fluorescent lights (which Mrs. Bean flicks on) buzz prominently on $\hat{5}$ (0:27:24) to create a dramatic dominant springboard from which the music can vault back in (when she leaves). After leaving Bean's cider cellar, Mr. Fox senses danger (the farmers hiding in a nearby bush), and the A-minor music halts suddenly on a V chord as he searches for the source of danger. The strings shimmer and swell like a rattlesnake on $\hat{5}$, as Mr. Fox's ears prick up and rotate like radar-detecting antennae, emitting a loud beeping sound on $\hat{7}$ (G♯), which forms the third of the V chord (0:30:38) (see Video 23 ✶).

The adorable *Benny and Joon* (1993) makes adorable use of a musicalized sound effect to illustrate the unique relationship between the neurodivergent Joon and Sam characters (Mary Stuart Masterson and Johnny Depp, respectively). When these two artistic freethinkers finally find themselves alone together (despite protective older brother Benny's [Aidan Quinn] efforts to keep them apart), the evening turns into a first date. They explore the chemistry that has been brewing between them, accompanied (nondiegetically) by composer Rachel Portman's quirky E-major underscoring featuring a descending-fifth sequence of dominant-seventh chords. Joon opens up by finally allowing Sam to use her art supplies (a real sign of trust), and they finger-paint intimately together. Sam reciprocates by making music with a balloon (0:51:46), by leaking air from the balloon to cause it to sing a warbling melody that is perfectly pitched and rhythmically coordinated to replace the melody line of Portman's theme (Figure 29) (see Video 24 ✶). This whimsically musicalized sound effect and its enigmatic interaction with the nondiegetic orchestral underscoring (with all the "fantastical gap" implications of this interaction) conveys the magic bubbling up between these two unconventional artists, who connect by sharing their art, but in their *own* idiosyncratic ways.

Atonement (2007) makes marvelous use of foregrounded typewriter sound effects as an integrated instrument in the nondiegetic music, to function as a "tool of narration" (Watts 2018: 28). These clicking typewriter sounds are not pitched, but they are rhythmically coordinated with Dario Marianelli's original underscoring to create a driving sense of motion—and even urgency, since they push slightly ahead of the musical beat (also noted in Watts [2018: 27]). The typewriter in this film is strongly associated with the central character, Briony (played by Saoirse

FIGURE 29. Portman's "Balloon" theme, with the melody line played first by the vibraphone, then by the balloon (ending with the popping of the balloon). *(Break-lines indicate omitted measures.)*

Ronan, Romola Garai, and Vanessa Redgrave, during her three stages of life), a determined young author whose later-in-life book the film recounts. The film depicts the events of Briony's book (likewise titled *Atonement*) as an ambiguous blend of flashback and fiction, which mirrors Briony's own difficulty in separating fact from fiction. The typewriter sound effects in this film likewise occupy an ambiguous space between diegetic and nondiegetic: they begin nondiegetically, accompanying the title cards of the opening credits, synchronized with the onscreen typing out of the film title (in a classic typewriter font).[10]

The typewriter sound becomes diegetic when the sound is visually anchored in the scene, synchronized with Briony typing at her typewriter. The typewriter sound becomes metadiegetic—and distinctively rhythmic, metrically aligned with

---

10. To explore the complex boundaries between diegetic, nondiegetic, and metadiegetic film sound, see Gorbman (1987), Stilwell (2007), Smith (2009), Neumeyer (2009), Winters (2010), and Heldt (2013).

the underscoring—as Briony begins moving determinedly about the house to announce the completion of her latest play to the household. The film makes this metadiegetic distinction clear by halting the typewriter sound whenever Briony approaches another human as she marches through the house (with the music likewise halting each time on a half cadence). This merging of diegetic sound into metadiegetic sound during the film's opening sequence is an important key for establishing how Briony merges reality into fiction, and the metadiegetic typewriter sound effect is used throughout the film to convey Briony's aggressive fictionalization of the world around her. When Briony or another character (Robbie, played by James McAvoy) is actually typing at a typewriter, the diegetic typewriter sound is a run-of-the-mill, unremarkable background sound. But when Briony's mind is at work fictionalizing, the metadiegetic typewriter sound is quite distinct: it is sonically foregrounded and rhythmically aligned with the 4/4 meter of the orchestral music in a striking and prominent manner.

In a particularly climactic moment that sets off a disastrous chain of events, Briony rushes to illicitly open Robbie's love letter to her older sister, accompanied by Marianelli's underscoring with typewriter soloist (0:27:57) (see Video 25 ✶). Both the music and the metadiegetic typewriter's rhythm ramp up as the tension in the scene swells (✶), the music ending with a *sforzando* attack of the typewriter when Briony reads the word "cunt." The four, *fff*, echoing typewriter attacks that follow, typing out the word *c-u-n-t* in Briony's mind, are pitched on $\hat{5}$ of the preceding C-major music (Puccini's "O Soave Fanciulla, O Dolce Viso" from *La Bohème* layered atop the Marianelli theme). The sound of that half cadence formed by the typewriter hangs dramatically in the air, like smoke from a fired gun, accentuating the awful significance of what Briony does next.

The musicalization and sonic foregrounding of the typewriter in *Atonement* elevates it to the level of the soloist in a typewriter concerto (a la Leroy Anderson), which draws into question the nondiegetic/metadiegetic function of the orchestral underscoring, and its role in Briony's narrative (as explored in Watts [2018]). The film's composer Dario Marianelli discusses his use of the typewriter in the orchestral underscore (Coleman 2017: 141 and 148), which likewise complicates the notion of the typewriter as *sound effect* versus *musical instrument*.

TRANSPOSED SOUND EFFECTS

Just as musical cues can be transposed to different keys for different narrative purposes, sound effects can likewise be transposed to different pitches to take on different meanings or coordinate with different cues. *The Grand Budapest Hotel* features a great example when Gustave and Zero reach Gabelmeister's Peak, and their aerial cable car stops midair to make a transfer with a monk. The music of the B♭-minor "Canto at Gabelmeister's Peak" theme temporarily drops out, leaving only

FIGURE 30. Melody lines formed by cable car pulleys.
a. Gustave's squeaking cable car pulley, pitched to form members of the V⁷ chord in B♭ minor (the concurrent key). The percussion in this passage is unpitched.
b. The monk's squeaking cable car pulley, using the same pitches but metrically displaced.

the percussion backbone, and the stylized sound of the cable car's squeaking pulley forming a kind of melody line (1:14:39). The pulley squeaks the pitches E♭-F-C, forming the chordal root, fifth, and seventh of the concurrent key's dominant-seventh chord (Figure 30a), and these squeaks are rhythmically synchronized with the meter. When the monk's cable car stops, his pulley squeaks the same pitches but metrically displaced (Figure 30b) (see also Video 26 ✱).

Moments later (during the same cue), when Gustave and Zero have reached the monastery and disembark from the cable car, its pulley squeaks out *new* pitches—D♭-F, which form the third and fifth of the tonic chord of the B♭-minor music (1:15:26). The sound effects of the squeaking pulleys in this scene punctuate the ascent up the mountain as if it were a conceptual "parallel interrupted period," a musical phrase structure that links two balanced halves with a midpoint pause (Figure 31). The theme in B♭ minor (and the cable car) stops briefly at the halfway point up the mountain, during which the pulleys squeak out dominant-chord pitches, forming a kind of "half cadence." This interruption of the symmetrical period structure heightens the sense of anticipation for its completion, and the theme (and cable car) resumes onward to resolve with an "authentic cadence" when the pulleys squeak out tonic-chord pitches.

In *Moonlight* the pervasive sound effects (buzzing fluorescent lights, door pounding, etc.) that roar quietly in the background of every scene are transposed

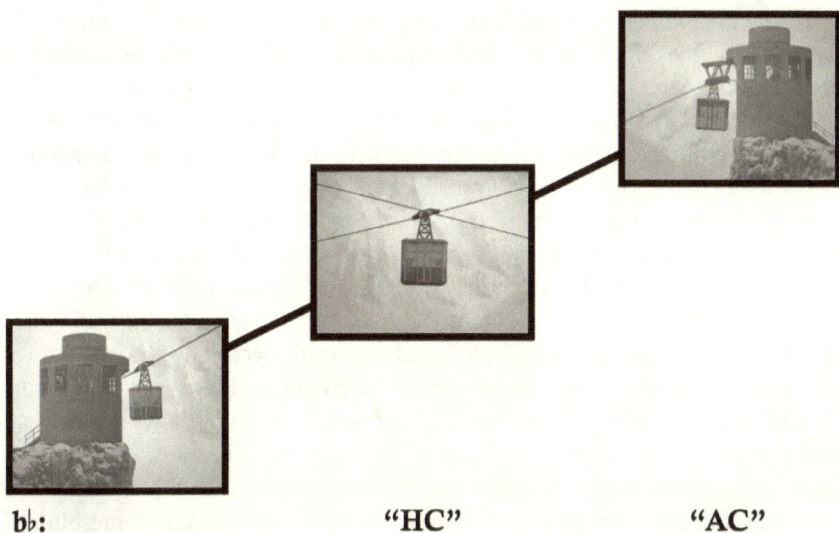

FIGURE 31. Conceptual "parallel interrupted period" structure of the journey up the mountain. The "antecedent phrase" ends with a "half cadence" at the midpoint, and the "consequent phrase" ends with an "authentic cadence" at the endpoint.

to different pitches to reflect the stages of Little's/Chiron's/Black's life. Crickets, for instance, are a ubiquitous background sound in many scenes, and their stridulatory singing is transposed to four different pitches throughout the film. In particular, there are numerous D- and B-pitched sound effects saturating the sonic ambience with these important tonics, corresponding to the D major and B major keys that associatively define the "Little" and "Chiron" (respectively) phases of his life. In chapter 3 we explored the transposition of both diegetic preexisting and nondiegetic original music in the different stages of *Moonlight*, and sound effects in this film are likewise transposed.

SIMPLY CONCORDANT SOUND EFFECTS

By far the most common tonal technique for pitched sound effects in film is one of simple concordance with preceding, concurrent, or ensuing musical cues. Once you start paying attention, you will be amazed at how often sound effects are pitched to $\hat{1}$, $\hat{3}$, and $\hat{5}$. But regardless of the scale degree they occupy within the key of a given musical context, *all* pitched sound effects interact with the tonality of the music around them, simply by virtue of being pitched *notes*. Be they chord tones or embellishing (non-chord) tones, we can hear and interpret their presence in the music as we would if they were produced by traditional "musical instruments."

In *The English Patient*, in addition to the associative C♯-pitched sound effects, there are countless other sound effects pitched to coordinate with the music. For instance, one evening in the Sahara, Katharine enthralls the expedition with a fireside performance of the tale of Candaules and Gyges, an ancient Greek story about a queen who orders her guard to kill her husband and marry her. Her narration of this story is peppered with significant looks between herself, her husband, and Almásy, alluding to parallels between the story and their own love-triangle. When Katharine utters the line "She shuddered," her words are echoed by the simultaneous shuddering rattle of a birdcall (0:28:34), concordant with the concurrent B-minor "Oriental Theme." As Figure 32a shows, this single-pitch birdcall enters on F♯, which matches the durationally and metrically accented principal note of the orchestra's ascending motives. When the ascending motives reach up to G, the birdcall pitch is transposed upward to a G as well. Figure 32b shows this birdcall repeated again after Katharine finishes her story with the phrase "The End" (0:29:23), but this time the birdcall remains solely on F♯, which is $\hat{5}$ of concurrent "Oriental Theme's" B-minor tonic chord. Figure 32c shows this same birdcall occurring one last time after Almásy ends his flashback of Katharine's story, before dozing off to sleep: the birdcall shudders on F♯ (0:30:06), which forms the root of the E-major "Passage of Time" theme's ii$^{ø4}_{3}$ chord, just before it resolves to I in a "Hollywood cadence."[11] And as the score reduction shows, the F♯ of the birdcall "resolves" to the E in the orchestral theme.

The *Wall Street* (1987) soundtrack may be famous for the Talking Heads "This Must Be the Place (Naive Melody)" montage celebrating the indulgent excess of 1980s new money, but there is an even more interesting musical sequence earlier in the film, when junior stockbroker Bud Fox (Charlie Sheen) is still trying to claw his way into the money game. David Byrne and Brian Eno's "America Is Waiting" plays during the tense scene in which Bud nervously watches the market fluctuate on the Terafly stock he sold to Wall Street legend Gordon Gekko (Michael Douglas) in order to prove himself (0:28:45). "America Is Waiting" is based around an E tonic (forgoing functional tonality in favor of *tonality by assertion*), and its melodic material consists of a relentless E-F-F♯ motive that never gets further than that F♯, reinforcing the restless sense of stasis and "waiting" referenced in the song title.[12] In the chaotic environment of the trading room during this sequence, numerous F♯-pitched phone rings are harmonically and rhythmically coordinated

---

11. For discussion of the "Hollywood Cadence," see Lehman (2013a).
12. "America Is Waiting" is structured as an ABABA five-part rondo, in which the A section centers on an E tonic, and the B section centers on a G tonic. The same thematic material is presented in both the A and B sections, so the G starting pitch of the B section does not feel like a completion of the hanging F♯ of the A section, because G immediately leads to G♯ and A (i.e., the two sections are modular and not elided).

FIGURE 32. Pitched birdcalls in *The English Patient*.

a. Birdcall pitches matching the upper line of Yared's "Oriental Theme."
b. Birdcall forming $\hat{5}$ of the "Oriental Theme's" final chord.
c. Birdcall interacting with the final measures of the "Passage of Time" theme, with the F♯ of the birdcall "resolving" to the E in the theme.

to interact tightly with the concurrent music (see a short excerpt in Video 27 ✴). The incessant F♯ phone rings drill down on Bud as he dejectedly waits and watches the Terafly stock flounder until the market closes. After the market has closed and the music has stopped, one more F♯ phone ring needles Bud as he closes his eyes in defeat (✴).

To demonstrate the sheer ubiquity of tonally coordinated sound effects, I have created a compilation video of $\hat{1}/\hat{3}/\hat{5}$ pitched sound effects from many films and television shows (see Video 28 ✴). This hodgepodge of films does not represent one particular time period or genre, in order to illustrate how pervasive $\hat{1}/\hat{3}/\hat{5}$ pitched sound effects are in Hollywood practice.

As I said before, the separation of music analysis and sound effect analysis into different chapters is a purely logistical one, to allow us to explore the theoretical apparatus of each in the book context. In the context of an actual film analysis, music and sounds are analyzed simultaneously as part of one integrated sonic fabric. (No doubt you noticed there were certain points in chapters 2–4 where I mentioned sound effects because it was simply impossible for me to separate them from the analysis of the music.) The following analysis of *Baby Driver* illustrates this.

BABY DRIVER (2017)

Techniques featured: associative tonality, transposition, tonal agency, intertextual tonality, parallel relationship, tragic-to-triumphant arc, functional tonality.[13]

RELEVANT PLOT SYNOPSIS

Baby (Ansel Elgort) is a getaway driver for crime boss Doc (Kevin Spacey) and his crew. Baby suffers from tinnitus and always has his iPod playing to drown out the din. A flashback reveals that his tinnitus started from a traumatic childhood car crash in which he lost his beloved mother (Sky Ferreira), who was a singer and who gave Baby his first iPod. One day after completing a job, Baby goes to a diner and falls in love with his waitress, Debora (Lily James), the two instantly bonding over their iPods and their love of music. Inspired by this new relationship, Baby begins to feel dissatisfied with his life of crime. He tries to quit, but Doc won't allow him to leave. Tensions among the crew rise as they embark on more dangerous jobs that go awry, and Bats (Jamie Foxx) and Buddy (Jon Hamm) grow increasingly antagonistic toward Baby. Buddy (who unfairly blames Baby for the death of his own girlfriend) tries to attack Baby and Debora, and they narrowly escape. The next day, Baby turns himself in at a police blockade and is sentenced to twenty-five years in prison. After five years as a model inmate, Baby is released on parole, and Debora is waiting for him, as they embark on their long-awaited road trip.

The first sound we hear in the *Baby Driver* soundtrack is a high-pitched D that rings in on the introduction of the Sony logo, subtly joined a few seconds later by strings, and persisting through the rest of the studio logo sequence. As soon as the logos fade away to reveal a busy city block, this D is matched by the D-pitched squeal of car brakes as a red Subaru pulls into the foreground. Before the driver of the car is even revealed, we see a close-up shot of his iPod as he cues up "Bellbottoms" by The Jon Spencer Blues Explosion—a song in D minor. As the annotated score

---

13. I first wrote this analysis of *Baby Driver* for Frank Lehman's edited volume, *Music Analysis and Film: Studying the Score*, which, due to the Covid pandemic, ended up being published *after* my own book.

FIGURE 33. Opening thirteen measures of "Bellbottoms" by The Jon Spencer Blues Explosion, with actions annotated.

excerpt in Figure 33 shows, the driver, Baby, is revealed on the first downbeat (D pitch, $\hat{1}$) of the song, his three crew members on the second, third, and fourth downbeats (also D's). And, preserving this musical pattern, they open the doors on the song's fifth downbeat, exiting the car amid a sea of Ds.

The hero of the 2017 bank heist film *Baby Driver* is a conductor with an odd kind of orchestra: armed with his iPod, getaway driver Baby entrains everyone and everything in his vicinity as instruments in the soundtrack of his life. With surrounding people and objects incorporated—both rhythmically and harmonically—into the music playing in his earbuds, Baby creates a living symphony out of car horns, gunshots, and passersby alike. In the opening sequence described earlier—the crew's first bank heist and ensuing getaway car chase—Baby cues up "Bellbottoms" to set the team in motion, and thereafter all their actions (and associated sound effects) are tightly synchronized to this music. Baby himself becomes animated (from his initially stoic pose) the moment the song's vocals begin, lip-syncing joyfully and enlisting his car's windshield wipers, steering wheel, and doors as percussion instruments as he watches the crew enter the bank. When a wailing police siren interrupts Baby's reverie, holding unnaturally on $\hat{3}$ (F), the song pauses and then restarts double time. With this new frenetic pace, things ramp up inside the bank, and gunshots are fired on the downbeats of the next four measures (now driven by the lead guitar). All these violent sound effects come to an abrupt halt when the song's instruments drop out for a spoken soliloquy by the lead singer, lip-synced by Baby.

Perfectly timed with the meter of the song and Baby's movements, the bank alarm begins wailing on $\hat{1}$-$\hat{3}$ (D-F) (0:03:04), soon joined by another alarm which beeps on $\hat{1}$ (D) as the rest of his crew piles back into the car. As they tear off from the scene of the crime, "Bellbottoms" veers into a frenzied instrumental interlude, supplemented with numerous brake squeals on $\hat{1}$ (D) and $\hat{1}$-$\hat{5}$ (D-A), a truck back-up beeper on $\hat{1}$ (D), and car horns on $\hat{3}$ (F) and $\hat{3}$-$\hat{5}$ (F-A)—plus countless precisely synchronized traffic sounds too numerous to enumerate. Collectively, all of these sound effects coordinate with the D-minor tonality of the "Bellbottoms" song to immerse the audience in the key of D minor for the first six minutes of the film—we'll delve into the significance of this key after we've explored the broader role of music in the film. The car chase ends when Baby pulls into a parking garage, where he and the crew hop out of his red Subaru Impreza WRX and into a green Toyota Corolla. Leaving his Subaru door open, the door chime pings on the quarter-note beat of the song, on a G pitch (0:06:05). The song moves to a G-major chord seconds later, and the door chime continues to ping quarter notes on $\hat{1}$ (G) as the song ends (in G major) and Baby chauffeurs the crew back to headquarters.

This type of tonally curated and meticulously choreographed musical space encompasses Baby wherever he goes; it facilitates his every action and is essential to his existence. Music is the driving force in his life, along with driving itself—he feels happiest when he is behind the wheel with his iPod. Music is Baby's activator (as we saw in the opening bank heist) as well as his soothing mechanism, and he uses music to process his experiences and make sense of the world. He frequently records conversation clips that he later samples and remixes to create song tracks, because music allows him to deal with people on his own terms. For instance, when a crew member ridicules him, Baby records the insult and reappropriates it, converting it into a form of musical expression that *he* controls. Music is the most vital part of Baby's identity, which is why we see his iPod before we even see his face. When he feels comfortable and in control of his life, every surrounding element, animate and inanimate, becomes perfectly pitched and timed to fit his music.

One of the clearest examples of this linkage of musicality with Baby's state of mind is the scene after the opening bank heist, when he goes on a coffee run for the crew (see Video 29 ✱). Swelling with confidence over the successful bank job, Baby joyfully grooves down the street to the coffee shop, listening to Bob & Earl's "Harlem Shuffle" (0:06:25). His surrounding environment grooves along with him, as a glorious host of car horns, crosswalk signals, ATM beeps, and construction sounds snap magnetically into the sonic landscape Baby creates and dances through. Even though "Harlem Shuffle" is strictly heard only inside Baby's earbuds, the musical collaboration between subjective and environmental sound feels entirely organic, as though the entire city is participating in the song. For example, Baby passes a young woman dancing on roller skates to the rhythm of "Harlem Shuffle," even though there is no visible music source near her. Words from the song lyrics are physically incorporated into the urban environment in the form of graffiti and stickers that Baby passes at precisely the moments those words are sung in the song.

A barista asks Baby a question at just the right moment to align his response with the "yeah, yeah, yeah" of the song lyrics. And as "Harlem Shuffle" approaches a climactic brass riff, Baby pauses in front of a music store, where a trumpet is suspended in the display window, positioned at *just* the right height for him to pose behind it and play air trumpet along with the riff. The fortuity of all the elements involved in Baby's sentient soundscape creates a fantastical realm that defies rational explanation. A series of "but how did that happen at exactly the right time?" type questions accumulate until they crumble under the suspension of disbelief that facilitates the unspoken contract between filmmaker and film viewer. After the opening car chase and coffee run scenes, the viewer realizes that this is no ordinary bank heist film, and that Baby is no ordinary protagonist; he is a quasi-superhero whose superpower consists of turning the world into his own personal opera.

Baby can't function without music, to the point where he sometimes has to stop his human instruments ("Wait wait—I gotta start the song over") and restart them ("Okay *go!*") in order to correctly choreograph a sequence of events. Or he'll partially rewind a song to account for unexpected complications and get the action back on track. When things aren't going well for Baby, actions and sound effects fall out of sync with his music. And the music in his earbuds becomes heavily muted (or disappears altogether) when he is in a situation that makes him feel trapped and incapacitated. Baby's enemies know that his well-being is strongly coupled with his musical coordination, and they use this against him. Antagonists attack him by invading his musical space, either by seizing control of sound effects, cutting off his music source, or misappropriating one of his signature songs. Buddy wages sonic warfare on Baby in increasingly aggressive stages. He begins early on by grabbing one of Baby's earbuds in a mock-playful gesture, and then yanking away both earbuds altogether. Later in the film, Buddy usurps the song Baby had specifically named as his favorite power anthem ("Brighton Rock" by Queen), blasting it in his car as he chases Baby down (0:55:28). Finally, Buddy shoots a gun past both of Baby's ears in a vicious attempt to burst his eardrums and take music away from him permanently.

But despite Buddy's attacks, Bats establishes himself as the *most* odious villain of the movie when he spews a vitriolic tirade against music—which, in the context of *this* narrative (and Baby's place within its construction), serves the sole purpose of conveying how utterly despicable he is. The crew boss, Doc, is far more subtle and nuanced in his villainy, but at the crucial moment when he makes a veiled threat toward Baby, he lifts his hand to authoritatively cue the chime of the elevator Baby awaits, symbolically demonstrating his control over Baby with his control of diegetic sound. This elevator chime cued by Doc is the same pitch (C) as the tension-filled tinnitus that fills Baby's ears when the boss makes his threat (0:52:25), imbuing Doc's sonic strike with an even more tangible impact.

Baby's tinnitus is the most prevalent audio motif of the film. The high-pitched D that opens the film is our first glimpse into the piercing din that plagues Baby—and this opening D literally *sets the tone* for the entire soundtrack (more on this later). Parsing this opening pitch a bit further, we realize that it is preceded by a faint dull roar (similar to the vacant sound heard when holding a seashell to one's ear), and the D's initial reverberating instrumental timbre subtly transforms into the clinical ringing-in-the-ears sound of tinnitus, before being masked by music with the onset of the strings. As the film progresses, we see that Baby's tinnitus and his need for music are intertwined, and we realize why music is so essential to Baby. A dramatic flashback reveals that Baby received his first iPod from his beloved mother as a child, learning to use it as an escape mechanism whenever his father became violent with them. He was listening to his iPod at the moment his fighting parents were killed in a car crash. The crash marked the onset of Baby's tinnitus, and he has used his iPod to block out this traumatic sound and memory ever since. Without music, the shrill ringing in his ears consumes him, and he feels lost and impotent. Baby's fervent need for music and concomitant impulse to drive thus stems from a desire to rectify this childhood trauma by staging the perfect run-through—striving to coordinate everything perfectly to achieve a successful do-over.

TABLE 19 Every occurrence of Baby's tinnitus in *Baby Driver*

Timestamp	Pitch	Relationship to Concurrent Music
0:00:05	D	$\hat{1}$
0:16:34	F	$\hat{1}$
0:21:55	F	$\hat{1}$
0:32:50	A♯	$\hat{1}$
0:34:08	G	$\hat{1}$
0:34:15	G♯	$\hat{1}$
0:48:10	D♯	$\hat{1}$
0:49:13	C	$\hat{1}$
0:52:22	C	$\hat{3}$
1:17:12	D♯	$\hat{1}$
1:18:08	G♯	$\hat{1}$
1:26:30	C	$\hat{1}$
1:28:48	F♯	$\hat{1}$
1:42:31	D	$\hat{1}$

Each time it occurs, Baby's tinnitus is pitched differently to fit the tonality of the situation. The tinnitus sound effect occurs fourteen times at eight different pitch levels during the film, serving as $\hat{1}$ of the concurrent music in all instances but one (Table 19). In the final instance the tone acts as $\hat{1}$ of the concurrent cue and $\hat{5}$ of the *ensuing* cue—an important moment that we discuss later. The pitch fluidity of his tinnitus is another manifestation of Baby's obsessive music drive, since even his hearing disorder is dynamically incorporated into the musical world he creates.[14] The tinnitus may be unbearable to him, but like all other sounds in his environment, Baby finds a way to turn it into music.

Baby's emancipation—both from the trauma of his past and the constraints of his present criminal job—begins when he meets Debora, his soulmate in music. They bond instantly through music when Debora breezes into the diner singing Carla Thomas's "B-A-B-Y," unknowingly spelling out his name as she walks past him (0:17:24). As they start a conversation about songs featuring their names, Debora and Baby bond over their shared feeling of being personally defined by music. The *source* of the music during their first three scenes together traces the trajectory of the relationship: they begin by listening to their own separate iPods in the diner, then they share earbuds to a single iPod at the laundromat (twirling around each other within earbud wire distance), and finally they create a sourceless music

---

14. In reality, tinnitus sufferers may occasionally experience pitch variations but not as frequently (or artistically) as Baby does in this film.

between them on their date at the restaurant. This dinner date features the Detroit Emeralds' song "Baby Let Me Take You," which begins the moment Baby shows up at Debora's house to pick her up (0:43:44) (see Video 30 ✱).

This song is intriguing in its apparent lack of source: it isn't strictly diegetic, because the foregrounded volume of the song and the complete lack of external sound effects imply that it is not playing over restaurant speakers (nor does Baby have his iPod playing). It isn't nondiegetic either, because of the marked way Baby and Debora interact with the music, both rhythmically and harmonically. As the annotated score excerpt in Figure 34 shows, their movements are perfectly synchronized with the musical nuances of "Baby Let Me Take You." Baby and Debora each run a finger around the rims of their wine glasses, generating $\hat{1}$ and $\hat{5}$ pitches on beats 1 and 3 (respectively). They clink their glasses together, producing a $\hat{1}$ pitch on the song's beat 2 (which is an accented beat in the funk genre). This type of musical correspondence marks every one of their physical motions and gestures—all while the cameras swirl tightly around them, simulating their earlier laundromat earbud dance. It becomes apparent during this sparkling moment that this music is metadiegetic (heard internally only by them), conjured by their powerful chemistry; their music is immediately backgrounded the moment another person (the waiter) interrupts their space, and it dies away altogether when Doc summons Baby across the restaurant. And it is through Debora's musical collaboration that Baby finds the strength to overcome his troubles and achieve his happy ending, as our examination of musical key demonstrates.

As noted earlier, *Baby Driver* begins in the key of D minor, a tonal center firmly established by the "Bellbottoms" song and the plethora of D-minor-centric sound effects during the opening bank heist. The film ends in the key of D major, so the tonal envelope of the film follows a tragic-to-triumphant arc. The key of D major occurs at the end of the film narrative, when Baby is released from prison to find Debora waiting for him. The song is Simon & Garfunkel's "Baby Driver"—the obvious inspiration for the title of the film itself—and it is immediately preceded by original orchestral underscoring by composer Steven Price, also in D major, played when Baby gets a postcard in prison from Debora (1:47:43). This D-major arrival is anticipated by a large-scale $\hat{3}$–$\hat{2}$–$\hat{1}$ motion in the preceding showdown scene between Buddy and Baby. In that climactic context a car door-ajar chime pings insistently on an F♯ pitch ($\hat{3}$) (1:40:27), followed by a sustained orchestral E pitch ($\hat{2}$), and finally an orchestral D ($\hat{1}$), matched by Baby's D-pitched tinnitus when Buddy blasts his eardrums.

But rather than leading *directly* to the tonally conclusive D-major "Baby Driver" song, this tonal trajectory is briefly interpolated by a crucial G-major event, hovering atop the sustained D pitch of Baby's tinnitus. The key of G major plays a pivotal role in the narrative. As previously mentioned, the persistent D—the last sonic manifestation of Baby's tinnitus in the film—functions simultaneously as $\hat{1}$

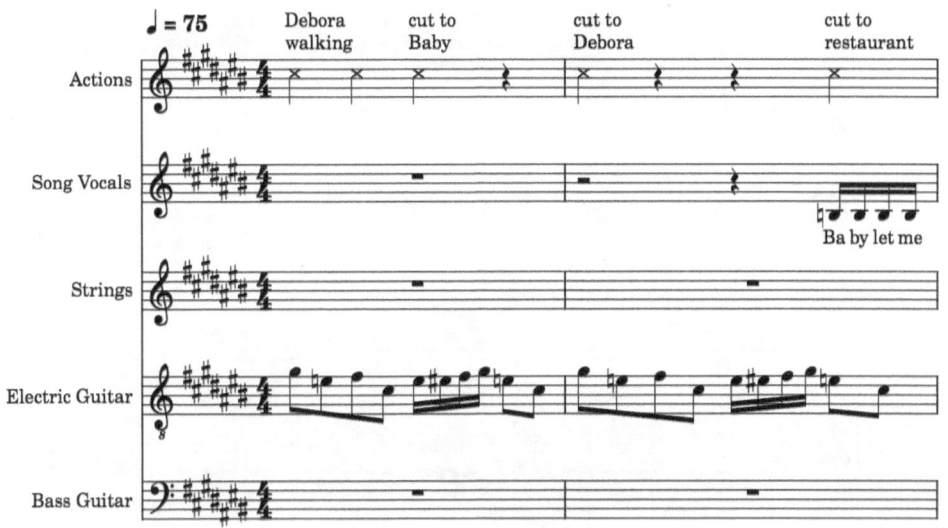

FIGURE 34. Opening fourteen measures of "Baby Let Me Take You" by The Detroit Emeralds, with actions annotated.

FIGURE 34. *(continued).*

FIGURE 34. *(continued)*.

FIGURE 34. *(continued)*.

of the concurrent music and $\hat{5}$ of the ensuing music, bridging the connection between D and G (the two most important tonal centers of the film). Baby and Debora join forces to vanquish Buddy, clearing the path to escape their former lives, and a glowing rendition of "Easy" in G major shines forth to celebrate their triumph (1:43:23). G major holds special significance for Baby, being the key in which both his beloved mother and his beloved Debora sing. The first time we encountered this key was during Baby's flashback of his mother, in which she was shown singing "Easy" in G major (0:33:10). The next instance occurs when Baby meets Debora and she sings the Beck song "Debra" for him, also in G major (0:36:50). When Baby and Debora defeat Buddy, we hear an orchestral version of "Easy" in G major (1:43:23), and when the couple begins their escape together, Debora plays a tape of Baby's mother singing the song in G major (1:43:35); and thus the two loves of his life are brought together by the key of G.

Within the film's overarching D minor-to-major tonal envelope, G major provides a large-scale plagal pathway to the achievement of D major. The plagal relationship between these keys evokes the redemptive associations of the so-called "Amen cadence" (IV–I). The plagal suggestion of salvation is especially fitting because the final (and lengthiest) instance of G major accompanies the montage sequence of Baby's court trial and prison sentence (1:43:35), during which Baby—dressed head-to-toe in white, and surrounded by white prison walls—receives absolution from every person affected by his criminal activities. This G-major sequence leads directly to the closing D-major "Baby Driver" song, granting our protagonist benediction to pursue his happy ending: Debora waiting for him beside a (white) car, with nothing but the open road before them. Thus the two women in Baby's life, represented by G major, deliver him from evil and help him find his way.

The song "Easy" was not originally written in G major, and it actually appears at a few different transposition levels in this film (Table 20). While we first hear "Easy" in G major (during the flashback of his mother), the original song was written and sung by The Commodores (1977) in A♭ major—which is the version we hear immediately after the G-major flashback version, when Baby cues it up on his iPod and walks away from the job, resolving to quit crime and run away with Debora (0:34:24). Later on, an orchestral version of "Easy" in C major accompanies Baby's road-trip escape fantasy (1:14:19). An orchestral G-major version plays over the mayhem when Debora and Baby defeat Buddy. And finally, Debora plays the tape of Baby's mom singing it in G major on their road trip, and throughout Baby's court trial and prison montage. Clearly, "Easy" is a cathartic piece of music for Baby, but the key of the song does make a difference in its effect on his life. The first G-based version we hear is what inspires Baby to realize he wants a different life. The A♭ version is powerful in its foregrounded volume level, but it doesn't actually lead Baby where he hopes—he *tries* to quit his crime job, but he is pulled back in against his will. During the C-major version Baby gets one step closer by seeing a vision of his new life, but the vision is dispelled and he remains stuck.[15] It is only when "Easy" is situated back in G major that Baby is finally able to escape. And, in light of its functional role within the overarching tonal progression of the film, G-major salvation is what allows Baby to achieve his D-major happiness.

"Easy" is not the only preexisting piece of music transposed to multiple tonal centers in this film: two other important songs are shifted into different keys, with meaningful effects on the narrative (Table 21). The first is "Harlem Shuffle," which is first sung in A minor by Bob & Earl (1963) and later transposed up to B♭ minor by The Foundations (1969). Baby plays both versions on his iPod as he goes on his post-heist coffee runs, but to very different effects. The A-minor "Harlem Shuffle" plays after the opening bank heist discussed earlier, when Baby is feeling great about his life and the successful job. But "Harlem Shuffle" plays in B♭ minor after a bank heist gone badly awry, when Baby is feeling frustrated and despondent (0:31:18). This B♭-minor version occurs after Baby has met the beguiling Debora

---

15. Taking the plagal key relationship even further, this C-major rendition of "Easy" provides plagal preparation for the G-major rendition. In narrative terms his C-major vision foretells the G-major reality that follows.

TABLE 20 Iterations of the song "Easy" throughout *Baby Driver*

Timestamp	Key	Plot Event
0:33:10	G major	Baby's flashback of mom
0:34:24	A♭ major	Baby walks away from job after flashback
1:14:19	C major	Baby's road trip fantasy
1:43:23	G major	Debora and Baby defeat Buddy
1:43:35	G major	Baby's court trial and prison montage

TABLE 21 Transposition of "Harlem Shuffle" and "Debora" in *Baby Driver*

Timestamp	Song	Key	Plot Event
0:06:25	"Harlem Shuffle"	A minor	Baby's coffee run, feeling *good*
0:31:18	"Harlem Shuffle"	B♭ minor	Baby's coffee run, feeling *bad*
0:37:41	"Debora"	B♭ major	Baby sings *alone*
0:39:07	"Debora"	A major	Debora and Baby listen *together*

and realized he wants a different life. Right after this B♭ minor, Baby goes into the diner to see his love interest, and he sings the T. Rex song "Debora" (1968) for her in the parallel major mode, B♭ major (0:37:41). Under Debora's restorative influence Baby transforms his mode from (B♭) minor to major.

And in fact, B♭ is subtly alluded to the first moment Baby lays eyes on Debora: during Bob & Earl's "Harlem Shuffle" in A minor, while Baby is in the coffee shop, the music temporarily moves to a ♭II chord [B♭ major] when his attention is caught by a young woman—Debora—walking past the coffee shop. The music moves back to a i chord as she walks out of sight and then modulates permanently to B♭ minor when Baby hurries out of the coffee shop to try and follow her. But she is nowhere to be seen, and Baby is mildly agitated as he looks up and down the street to find this mysterious girl, while the song fades out in the key of B♭ minor. Thus, even before he has officially met her, Baby is perturbed from his A-minor tranquility to B♭ minor when Debora first walks into (or past, more accurately) his life.

Immediately after this B♭ minor-to-major modal transformation involving T. Rex's "Debora," Baby and Debora jubilantly listen to this song together on his iPod, now in the key of A major (transposed from the recording's original B♭ major; more about this transposition in chapter 6). So whereas Baby's initial feel-good song was in the melancholic key of A minor, Debora helps Baby not only return to the A tonic but transform the overriding mode from minor to major. We can reconstruct a rough narrative trajectory from these transpositions: Baby starts out feeling fine in A minor

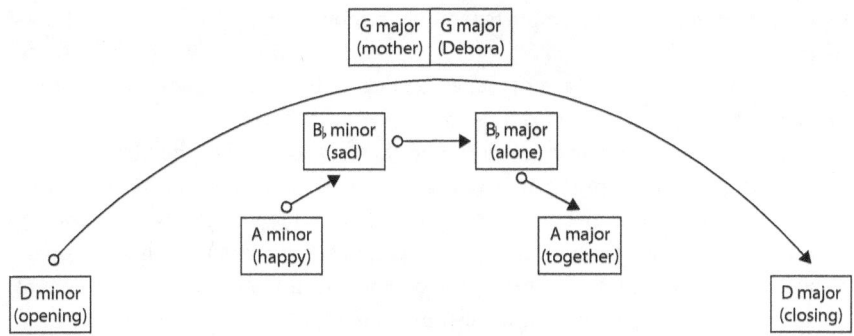

FIGURE 35. Large-scale tonal relationships across *Baby Driver*.

but becomes unsettled in B♭ minor; Debora then helps him turn that B♭ minor into B♭ major when he is singing *alone*, and the two of them listening *together* turn B♭ major into A major. There is thus a smaller-scale tragic-to-triumphant arc embedded within the larger one (D minor to major). Figure 35 captures the important tonal relationships in this film.

It seems likely that the T. Rex "Debora" song was the inspiration for the Debora character's name, just as Simon & Garfunkel's "Baby Driver" was the inspiration for the Baby character's name, as well as the name of the film itself (director Edgar Wright has discussed his music-driven creative process for this film in interviews, as explored in chapter 6). That the main characters and the film title are built around preexisting songs demonstrates how fundamental music is to this film—at an even deeper level than the obvious role of music in the narrative. As a thought experiment in reconstructing the development of this soundtrack, one can imagine that because the "Baby Driver" song was preselected as the ultimate musical payoff of the film, its key (D major) served as the tonal germ of the soundtrack. The D-minor starting-point ("Bellbottoms") would naturally come next, to create a satisfying tonal arc leading to the D-major ending point. And the D-pitched tinnitus that opens the film plants the seed for the D minor-to-major arc—*setting the tone* for the whole soundtrack, as I punned earlier. G major would be the next step, to set up the long-range plagal atonement for Baby's sins, with the G-major version of the "Easy" song produced specially for the film.

Incidentally, the D–G–D relationship features prominently in both the opening "Bellbottoms" song (which begins in D and ends in G, as mentioned earlier) and the closing "Baby Driver" song (in which the intro and most of the verse consists mainly of I–IV–I progressions). Finally, the smaller, internal arc of A minor-to-major fits nicely within the overarching D minor-to-major arc. Together, these main keys create a conceptual sort of meta-progression across the film, with the i→I frame filled in with important contributions from V (A major) and IV (G major).

Music in *Baby Driver* was not tacked on during post-production to fill silences or make chase scenes more exciting: as we see in chapter 6, music was the *progenitor* of the narrative, the characters, and the actions—and of the film itself. The film's protagonist uses music to align the world around him and direct his life according to the score of his choosing. But despite his superhero-like power to entrain and conduct everything and everyone as instruments in his life opera, Baby is stuck in a criminal life he cannot escape. Meeting Debora makes him realize he wants a different life, but his criminal colleagues manipulate him, at times musically, to keep him in place. But he "gets by with a little help from his friends," as the two women he loves give him the tonal boost he needs to drive off into the sunset in pursuit of his happy ending.

Hailed as a "car-chase opera" (White 2017), *Baby Driver* features a truly integrated soundtrack in which music, sound effects, and dialogue are woven together to form a living musical fabric that occupies *the* central role in the narrative. Music and sound in this film do not simply decorate empty space and overdetermine the visuals; turning off the audio and watching this film with closed captioning would strip away crucial information about Baby's character and narrative arc. Key tells a story in this film—whether this tonal story-beneath-the-story was consciously planned or one of the "happy accidents" (Hall 2017; more on this quote in chapter 6) of the production process.

CLOSING THOUGHTS

As legendary sound designer Walter Murch says, sound is not "something that you can only sprinkle over a film at the end of the process, but it's a force that can be used from the beginning in the telling of the story" (Murch 1995: 244; cited in Kulezic-Wilson 2020: 13). In this chapter we have seen how sound effects integrate with the music to create complex tonal textures that transcend the diegetic-nondiegetic boundary. In recent decades scholars have been challenging the boundaries of Hollywood's traditionally compartmentalized soundtrack: among many others, Rick Altman, McGraw Jones, and Sonia Tatroe (2000) coin the term *mise-en-bande* to depict the soundtrack as a unified entity; James Buhler (2001: 39–40) describes music as a "subsidiary element in the sound design" and encourages scholars to consider the "unified soundtrack"; David Neumeyer (2015) advocates for analyzing the unified soundtrack; Sergi Casanelles (2016) blurs the conventional definitions of sound and music by showing how even traditional-sounding orchestral instrumentation is often digitally constructed and/or enhanced; and Danijela Kulezic-Wilson (2020: 6) illustrates how technology is continually closing the "ever-diminishing gap between what is perceived as music and what is perceived as sound effect."[16]

But thus far, when scholars talk about hearing sound effects musically, they mostly speak in terms of rhythmic, timbral, functional, and logistical coordination between sound and music, *not* the harmonic coordination of sound effects' specific pitches in relation to the keys of the music surrounding them. My work in this chapter takes the concept of music-sound integration to the next logical step, by proposing a new way of hearing film that includes listening to how the *pitches* of sound effects fit within their sonic environment. This new approach will likely cause you to start paying tonal attention to the sounds in your everyday life—the harmony of your daily spheres, so to speak. The whir of your dishwasher, the beep of your microwave, the vibration of your phone—all of these sounds carry pitches and can be heard to interact with other sounds and music in your environment.

Even *without* considering the overarching tonal design of a film, one can do micro-interpretations of sound effects within a localized context. For example, at the end of *Pretty Woman* (1990), when Edward (Richard Gere) goes to rescue Vivian (Julia Roberts) on his white horse (limousine), the tender underscoring builds up to a IV chord, which resolves to I (just as Edward smiles), at which point Vivian is shown sitting by the window of her tower (apartment) with church bells in the distance pealing on $\hat{1}$-$\hat{2}$-$\hat{1}$—which both affirms the redemptive qualities of the plagal cadence and evokes wedding bells (see Video 31 ✷) (1:54:00). In the iconic shower scene of *Psycho* (1960), when Marion's dying hand grasps the shower wall as she slumps limply into the tub (0:48:00), Bernard Herrmann's grisly scoring begins with a sharp F-E motive barked by the cellos and basses; when the camera zooms in to reveal the water circling down the drain hole (forty seconds later), the sound of the water swirling past the drain's edge insistently reprises those ominous F and E pitches. In *Annie Hall* (1977), when Annie (Diane Keaton) sings "It Had To Be You" in a nightclub, a phone rings loudly as one of several comical interruptions during her song; the phone ring's pitches ($\hat{1}$-$\hat{3}$) function as the chordal seventh and ninth of the music's concurrent ii^7 harmony, spurring it on to its dominant half-cadence resolution (see Video 32 ✷) (0:34:13). In *30 Rock* episode 5.14 (2011), when an airplane passenger assures Liz Lemon (Tina Fey) that he is *not* an air marshal—which he absolutely is—the music hangs on a V^7 chord waiting to resolve to I, while the airplane's announcement chime rings on $\flat\hat{7}$ and $\flat\hat{5}$. These two pitches, both blue notes, create a comical "wah wah" kind of sound, highlighting the humor of the moment (see Video 33 ✷).

---

16. For an excellent overview of the historical segregation and merging of soundtrack elements, see Kulezic-Wilson (2020: chapter 1) and Casanelles (2016).

Let's not forget about pitched dialogue, such as the poignant moment in *I Love You Phillip Morris* (2009), when a seemingly dying Steven (Jim Carrey) imagines hearing his mother call his name on pitches ♭$\hat{7}$-$\hat{6}$, mimicking the ♭$\hat{7}$-$\hat{6}$ of the Dorian music's thematic line during that scene (see Video 34 ✶) (1:21:10). In *Bridesmaids* (2011), Annie (Kristen Wiig) listens to Lillian's voicemail message in which she says, "Hey it's Lillian, leave a message after the *beep*," where the word "beep" is beautifully delivered as $\hat{5}$ (and rhythmically aligned as beat 1) of the concurrent music (1:15:25). And even without the immediate presence of music, one can still interpret the pitches of sound effects in relation to one another—for instance, in the chilling scene of *2001: A Space Odyssey* (1968) when the HAL computer begins killing off crew members by cutting off their life support systems, the two different machine beeps form a tritone, the age-old Western sonic signature of danger (see Video 35 ✶) (1:36:20).

Of course, there is no *requirement* that the pitch of every sound effect should necessarily hold some sort of tonal "meaning": it's simply about letting sounds become music. Sound effects pitched at $\hat{1}$, $\hat{3}$, or $\hat{5}$ (of whatever music is playing concurrently) will obviously nestle into tonal place without any ambiguity; but *every* pitched sound effect can be heard to interact with and contribute to the tonality of music in some way, regardless of its scale degree. With a $\hat{6}$-pitched sound effect, allow your ear to turn the concurrent music's tonic chord into a I$^{\text{add6}}$ or vi^7 harmony; a ♭$\hat{7}$ sound effect can jazz up the music's tonic or destabilize it into a V^7/IV; and a ♯$\hat{4}$ sound effect can introduce the Lydian sense of wonderment[17] or threaten the music's otherwise seemingly comfortable tonic harmony with tritone implications—all dependent on context, of course. We are now consciously hearing previously "unheard" sound effects.

Chapter 6 explores questions of intentionality in film tonality, along with a brief nod to tonal design in other media, before drawing conclusions about the tonal analysis of film.

---

17. See Lehman (2012b: 14–18 and 31–32) for an account of Hollywood's use of the Lydian mode in producing a sense of "soaring wonderment" in film.

6

Happy Accidents: Intentionality and Other Closing Thoughts

TO KEY OR NOT TO KEY: THAT IS THE QUESTION (OF INTENTIONALITY)

"Did they do it on purpose?" This is a question that comes to many people's minds when they first hear about film tonality. As discussed in chapter 1, intentionality is a thorny issue to nail down, especially in the context of multi-artist works like film. But we can use several factors to form a better understanding of how a tonally designed soundtrack might come to be. First and foremost, music is of paramount importance to some directors, and the effect of their meticulous attention to the musical development of their films cannot be understated. In her discussion of *mélomanes*, or musical auteurs, Claudia Gorbman includes Quentin Tarantino, Woody Allen, Spike Lee, Martin Scorsese, and Stanley Kubrick (among others) as directors for whom music is a central "authorial signature" (2007: 151). We can certainly add some of the directors we have explored in this book to the list of well-established *mélomanes*.

Anthony Minghella (director of *The Talented Mr. Ripley*, *The English Patient*, and *Breaking and Entering*) begins planning out the music for his films from the earliest stages of development. Based on his extensive interviews, Mario Falsetto describes music as "Minghella's first love" and the factor that first sparked his interest in words at a young age (2013: xxiii). In Minghella's own words, "music is at the center of every film that I've ever made," and he describes his process as one in which "everything grew out of the music" (Falsetto 2013: 18, 133).[1] Minghella

---

1. This second quote is taken from Nick Taylor's interview with Minghella (titled "He Shoots, He Scores") in the *Guardian* on March 11, 2005, which Falsetto reprints in his 2013 book.

works closely with composer Gabriel Yared in crafting his films, and "the relationship with Yared was never a conventional director-composer one. Yared was involved with each film early in the production process, sometimes even at the script stage" (Falsetto 2013: xxiv). For example, in discussing *The English Patient*, Minghella states: "When I'm writing I'm already marinating the film in some kind of musical landscape, and I bring those elements to [composer] Gabriel [Yared] as a gift. So for instance, in *The English Patient*, I came to him with a sense of how Bach was a very strong ingredient" (Bernard and Khanna 2007).

Wes Anderson's (director of *The Grand Budapest Hotel*, *The Royal Tenenbaums*, *Fantastic Mr. Fox*, *The Darjeeling Limited*, and *Moonrise Kingdom*) status as a *mélomane* of the first order is as undisputed as his auteur status, and scholars frequently name Anderson in discussions of music-driven directors (see, for instance, Winters 2012, McQuiston 2017, and O'Meara 2014). In the context of *The Royal Tenenbaums*, for example, music supervisor Randall Poster (who has supervised music in all of Anderson's films except for his first, *Bottle Rocket* [1996]) says that "Anderson called me to discuss music as he was putting the script together" and "we had seventy-five percent of the songs picked out and licensed before we even started shooting the movie" (Kubernik 2006: 196 and 112–113, respectively). Ben Winters points out that Anderson fills his films with characters who play musical recordings as a focal form of self-definition—characters who are "composers of their own scores" (2012: 46)—as a way of mirroring Anderson's own musical aesthetic, music curation process, and reverential engagement with music.

Edgar Wright (director of *Baby Driver*) has stated explicitly that music inspired this film. Long before the film began to take shape in his mind, certain songs suggested imagery and action sequences to him: "I would literally listen to 'Bellbottoms' by the Jon Spencer Blues Explosion and visualize this car chase and be able to visualize: 'So the first half of this section, they're all pulling up. On these guitar stabs, you cut around the gang and they get out and go inside. And now he's on his own and he starts singing along with the song. And then they come back and it's building up to the song really kicking into gear and then they drive off'" (Hall 2017).

It wasn't until decades later that Wright began envisioning "a car chase movie driven by music" and "started to get the idea of a getaway driver who listens to music the whole time" (Hall 2017). In 2002, Wright "road test[ed] the idea for the first scene" of *Baby Driver* in a three-minute music video he directed for "Blue Song" by Mint Royale—likewise featuring a getaway driver waiting for his colleagues to rob a bank, but with far less fastidious synchronization than he would incorporate in *Baby Driver* fifteen years later (Hall 2017).[2] Wright describes *Baby Driver* as "an action car chase movie that is completely powered by music, so it's

---

2. Wright's "Blue Song" video can be viewed here: www.youtube.com/watch?v=iHbndkcLM6A&ab_channel=EdgarWrightHere.

almost like an action-musical" (Collis 2017). He explains the synchronization in this film as a precisely "mathematical" process with "a lot of to-the-millisecond planning," staging the action both forward and backward to work between a multitude of anchor points in every song (Hall 2017). Wright takes a music-driven approach to his other films as well, such as *Scott Pilgrim vs. the World* (2010) and *Shaun of the Dead* (2004).

It is clear from these brief vignettes that, despite typical Hollywood workflows that tack music on in postproduction, some directors treat music as a crucial structural element in their films. But even with a *mélomane* at the helm, we cannot assume that every musical element we can observe and interpret in a film is deliberately planned. Wright speaks of the "happy accidents" that arise during filming and editing, in which unpremeditated phenomena contribute to the narrative in unexpected but fortuitous ways (Hall 2017). For instance, one adventitious result of Wright's relentless forward-and-backward working process in *Baby Driver* was that this process "also speaks to the [Baby] character," since "when he's rewinding the song, he's trying to get back in control again" (Hall 2017). *Thus not every meaningful element in a film soundtrack was intentionally or consciously premeditated for that purpose.*

As stated in chapter 1, a multifaceted, multi-artist production like film acquires a gestalt that is greater than the sum of its individual contributions—and even consummate auteurs and *mélomanes* acknowledge this synergy. Wright, in discussing the exquisitely choreographed composition of his film, states enthusiastically that "it is all about working together. I know that sounds like stating the obvious, but there are lots of other movies where the different teams are all working in isolation. I like to do the opposite" (Collis 2017). Wes Anderson's "desire to 'deauthorize' himself by situating his creative contributions in a more collaborative framework" has been documented by Winters (2012: 45).[3] Even in the context of early classic Hollywood films, Nathan Platte notes that "in its final form, the soundtrack to *Spellbound* represented not a single or even shared vision, but rather an intricate conglomeration of ideas, revisions, and interpolations" (2011: 457).

In teasing out the concept of intentionality in the arena of film tonality, there are slightly different considerations for original and preexisting music. For original music it's easy to understand that composers choose the keys in which they compose cues. For example, Ennio Morricone (who scored virtually all of Sergio Leone's films) composed the entire score for the films *A Fistful of Dollars* (1964) and *For a Few Dollars More* (1965) in the key of D minor. Morricone remarked: "When I begin a theme in a certain key, say D minor, I never depart from this

---

3. Pascal Rudolph (2022) proposes the name "Musical Idea Work Group" to capture the collaborative working environment of soundtrack design and explore the realities of the creative process behind the curtain of the "auteur *mélomane*" designation. His discussion presents Lars von Trier as a case study of contemporary filmmakers crafting soundtracks from preexisting music.

original key. If it begins in D minor, it ends in D minor" (Fagen 1989: 106). And Max Steiner, in his compositional sketches, specified the pitches of sound effects and even dialogue (such as whistling or humming) to coordinate with the tonality of his musical scoring (✶).[4] But even without direct composer commentary, key intentionality can sometimes be quite transparent to deduce. Gabriel Yared, for instance, has not commented directly on key, but it's easy to infer that he based the keys of his original themes on the keys of the preexisting works that anchor those soundtracks (for example, Bach's *Goldberg Variations* engendering the key of G major in *The English Patient*)—especially since these preexisting works were often presented to him by Anthony Minghella before the projects began and especially because we can see this same workflow in multiple Minghella-Yared films.

Transposition is one of the clearest indicators of tonal intentionality, because a musical cue/recording doesn't just turn up in a different key without somebody moving it there. In the film *Moonlight*, for instance, the many transpositions flagged my attention as noteworthy, and I formed an interpretation based on their interaction with the narrative (in chapter 3). Specific knowledge of how or why the filmmakers transposed this music was not needed in order to form this interpretation; but for the sake of argument, let us see what would happen if we did take this information into account. As mentioned earlier, *Moonlight* features a number of preexisting hip hop songs with lowered pitch, and these were all transposed downward (for the film) via the "chopped and screwed" technique. This technique entails slowing down songs to lower their pitch (as well as remixing them) to create a hazy, sedated, dreamlike atmosphere in the music. Having emerged in the early 1990s hip hop scene of the urban South, the chopped and screwed technique fits naturally in the setting of *Moonlight*, being appropriate both to the style of music and the time period and location of the film.

Inspired by director Barry Jenkins's use of chopped and screwed music, composer Nicholas Britell decided to apply this technique to the theme he composed for the film. Britell wrote and recorded "Little's Theme" in D major, then chopped and screwed it down to B major for "Chiron's Theme." For "Black's Theme," he reorchestrated the melody for cello (rather than violin) but still recorded it in D major, chopping and screwing it all the way down to A major (Cooper 2017). In an interview Britell says that he used transposition to "provide a sense of cohesion across chapters while also allowing for transformation" (Shapiro 2017). But would Britell have thought to compose a transposing theme if Jenkins hadn't already established chopped-and-screwed as part of the film's ethos? Perhaps this is one of the "happy accidents" Edgar Wright spoke of, that Britell was able to harness this

---

4. My deepest gratitude to Jeff Lyon and Brent Yorgason for telling me about (and sending me) these Steiner sketches (✶), as they meticulously analyze and digitally catalog the Max Steiner Collection, housed in the Film Music Archives of the Harold B. Lee Library at Brigham Young University.

transposition technique in service of the narrative? As Britell freely admits, "one of the amazing things about the film scoring experience is that I think you really never know how things are going to turn out or work" (Shapiro 2017). Britell speaks about his cues in terms of their keys—D major, B major, A major—even though the "B-major" and "A-major" themes don't *technically* exist in those keys (Britell did not write scores containing those key signatures, and no musicians performed in those keys). But these keys *do* exist for the film viewer-listener to interpret—and we can interpret anything we see/hear in a film regardless of the logistical or technical reasons it exists. If Britell, in response to the reporter's question, had sheepishly replied that the transposition of his theme was the accidental result of a sound engineer setting their sandwich down on the computer keyboard while running Audacity, it wouldn't invalidate the correlation between the lowering theme and Little's/Chiron's/Black's deepening voice or render it any less compelling.

Obviously, composers can choose the keys in which they compose; but even with preexisting music and sound effects, filmmakers today have music/sound libraries with virtually infinite selections to choose from—as well as sophisticated software that can instantly alter and transpose pitch levels. So one can effortlessly put any music/sound into any key/pitch. In *Baby Driver*, for instance, we saw (in chapter 5) that T. Rex's 1968 "Debora" recording was transposed down from B♭ to A major for the film. In general, one might assume that a lower-pitched recording is a by-product of slowing down the song speed (such as we saw in *Moonlight*). However, the tempo of this song was actually *sped up* for the film (from ~78 bpm to ~84 bpm), which means that the music/sound editors had to alter the pitch (key) separately and purposely, in the opposite direction. Likewise in *Baby Driver*, Simon & Garfunkel's 1970 recording of "Baby Driver" is transposed downward. Their original recording is pitched slightly higher (~600 Hz) than the standard modern D pitch (~587 Hz)—sounding somewhere between D and D♯—even though the song was performed in D major, as can be verified from Simon & Garfunkel's live performances. (Presumably, the original recording was sped up by the studio sound engineers for LP timing reasons.) The film, however, adjusts the song down to the standard D pitch (without adjusting its speed), which highlights the significance of the D-major key in *Baby Driver* and makes the tonal framework of the film seem even more deliberate—*if* one considers deliberateness as a metric of legitimacy for analytical interpretation.

However, as much as we might wonder about intentionality in art, there is often no way to *definitively* delineate how much of art is consciously planned and how much results from unconscious creative genius, in the form of "happy accidents." Take, for instance, Jackson Pollock: in the late 1990s, scientists discovered that Pollock painted in fractals (Taylor, Micolich, and Jonas 1999). If an (anachronistic) interviewer were to ask Pollock if he purposely painted in fractal form, and Pollock bristled that he did not, would that invalidate the presence of fractals (or our amazement in them)? And if Pollock responded that he *did* paint fractals intentionally, would that serve as definitive proof? (That is, could we be certain there was no element of aggrandizement or self-mythography, as Stravinsky and many other artists are prone to, in their self-analyses?) Given Pollock's dynamic "drip technique" method, it seems virtually impossible that one could *purposely* set out to paint in fractal form. The bottom line is, fractals can be found in Pollock's work, and I would argue that it's far more fascinating that he *didn't* paint them consciously. Likewise, the "Society of the Crossed Keys" key occurs at the Golden Ratio point of *The Grand Budapest Hotel*, and whether Wes Anderson[5] did this "knowingly" or not, his work joins the *long* list of human creations, such as the Great Pyramid of Giza, which adhere to the Golden Ratio either by "accident or design" (as carefully worded by David Burton in his history of mathematics textbook [2011: 60]).

Even the vast networks of symbolism in Coen brothers films may be deemed as "happy accidents" if we abide by filmmaker commentary, since Joel Coen (in response to the many beloved interpretations of symbolism in *Barton Fink* [1991]) stated: "We never, ever go into our films with anything like that in mind. There's never anything approaching that kind of specific intellectual breakdown. It's always a bunch of instinctive things that feel right, for whatever reason" (Coen and Coen 2006: 94). The point is, interpretation is the realm of the reader, and an analysis is convincing because of its ability to enhance our engagement with a work and communicate to other readers. As Frank Lehman says, "To limit ourselves to only poietic support in analysis of film tonality would be self-defeating: we artificially constrain the kinds of claims possible to a small (or, likely, nonexistent) interpretive universe dictated by the composer him or herself" (2012b: 79–80n10). According to W.K. Wimsatt and M.C. Beardsley, an artwork "is detached from the author at birth and goes about the world beyond his power" and "belongs to the public" for interpretation, and "critical inquiries are not settled by consulting the oracle" (1946: 470, 487). And because the very concept of authorship is a synecdochic construct (as discussed in chapter 1), there is no knowing which of the scores of oracles working on a film soundtrack one should consult.

TONAL CONTINUITY IN OTHER MEDIA

Other large-scale, multipartite media formats utilize their own forms of tonal design. For instance, in the context of radio and club playlists, "harmonic mixing" is a method of continuity-editing to ensuring smooth tonal transitions (sometimes called "tonal flow") between contiguous cues. Software programs such as Mixed in Key and Rapid Revolution help disc jockeys (both radio and club) design playlists based on track properties such as key, among others.

---

5. By "Wes Anderson," I do of course mean the synecdochically constructed "author" figure.

Video games use tonal design to coordinate the interaction of sound effects with music.[6] For instance, in the classic Nintendo *Super Mario Bros* (1985) the vast majority of the music is in C major (except the "Underworld" theme, which approximates C Dorian, and the "Castle" theme, which is an atonal collection of diminished harmonies), and most of the sound effects are pitched to fit the key of C. The "fireball" sounds on $\hat{5}$, and this is the most ubiquitous sound effect, as players spend most of their game time dispensing fireballs and pressing the [B] button to run at top speed. The "1-up mushroom" sound effect forms a pert $\hat{3}$-$\hat{5}$-$\hat{3}$-$\hat{1}$-$\hat{2}$-$\hat{5}$ melody line. The "super mushroom" sound consists of $\hat{5}$-$\flat\hat{6}$-$\flat\hat{7}$, which alludes to the Aeolian cadences of the main "Overworld" theme and the "Course Clear" fanfare. (I also point out that the "Castle Fanfare" after defeating each Bowser-imposter ends with a half cadence, spurring Mario on to the next level, as a motivational form of cadential frustration. Only once he defeats the true Bowser and rescues Princess Toadstool is a triumphant PAC attained.)

Even in the context of software/app design, pitch matters. For instance, the Google Chat computer app's default notification sound effect (called "Tones") is a G-G octave, which forms a pleasant dominant or tonic *bing-bong* against any C-based or G-based music you are listening to at your computer as you work and chat. I'm always surprised at how often my Google Chat chimes in harmoniously with my concurrent music, and I speculate that the G pitch was chosen because C major and G major are such ubiquitous keys—the most commonly used keys in popular music, in fact.[7] Thus G is the likeliest pitch to harmonize euphoniously with the music around you. All of the other clearly pitched notification sounds employed by Google Chat are G-major-centric as well (Table 22).

From our perspective as citizen-analysts, it doesn't matter whether a Google engineer made these choices consciously (with the awareness of key distribution in popular music) or unconsciously (because the G pitch sounded most appealing to that engineer based on the music playing on *their* desktop at the time). Because tonal design studies are not bound by authorial intent, the result *is what it is*—whether "deliberate or adventitious," to quote Hans Keller's pithy phrase once more (1955: 85). And along those lines, notice that the use of G major for Google constitutes a tonal wink (akin to NBC's tonal wink, mentioned in chapter 1)!

---

6. Elizabeth Medina-Gray discusses pitch as one of the parameters of smoothness/disjunction in video game music (2019) and discusses consonance in video games (2021: 186–191).

7. Dave Carlton (2012) analyzed thirteen hundred popular songs, and Eliot Van Buskirk (2015, along with Kenny Ning) analyzed the entire catalog of thirty million songs on Spotify, and both arrived at C major and G major as the most popular keys. As both Carlton and Van Buskirk have pointed out, these two keys—in addition to having simple key signatures—are the most *physically convenient* when playing the piano/keyboard (C major) and the guitar (G major), the most commonly used instruments in popular Western music.

TABLE 22 Google Chat's pitched sound effects for notifications

Sound Effect	Pitch	Role
"Tones"	G/G	$\hat{1}/\hat{1}$
"Welcome"	G/E—G/D	$\hat{1}/\hat{6}$—$\hat{1}/\hat{5}$
"Nudge"	G/E—B—G/D	$\hat{1}/\hat{6}$—$\hat{3}$—$\hat{1}/\hat{5}$
"Sweet"	G/B	$\hat{1}/\hat{3}$
"Music Box"	B	$\hat{3}$
"Calm"	G/B	$\hat{1}/\hat{3}$
"Treasure"	F/A/C—G/B/D	♭$\hat{7}$/$\hat{2}$/$\hat{4}$—$\hat{1}$/$\hat{3}$/$\hat{5}$
"Piggyback"	D/B—G/D	$\hat{5}$/$\hat{3}$—$\hat{1}$/$\hat{5}$
"Shrink Ray"	D/A/D	$\hat{5}$/$\hat{2}$/$\hat{5}$

NOTE: The sound pitches, as of December 2022, make reference to the key of G major—the tonic triad specifically, for eight out of the nine ("Shrink Ray" outlines the dominant triad). Notice that "Treasure" forms a partial Aeolian cadence (♭VII-I).

There is one subtle but important difference in the logistical motivation for tonal design in film versus radio and club sets. In radio and club sets, tonal design is used in the context of a continuous stretch of uninterrupted music, in which one song directly abuts (or mixes into) the next—such that lack of tonal continuity will be starkly apparent, and perhaps even jarring. In these types of settings, music is foregrounded—it is the focal aural activity presented to the audience. Even if listeners are engaged in an ancillary activity (e.g., working while listening to a radio set, or dancing while listening to a club set), the experience that is presented to them is aurally focused on this one activity. A film soundtrack, however, presents the audience with multiple levels of aural activity, and music is not usually the main sonic event. Most often in film, dialogue is the foregrounded aural activity on which the listener is focused, and music is of secondary or tertiary importance after dialogue and sound effects. And in addition to being backgrounded, the musical texture of a film soundtrack can be disjuncted by minutes of non-musical time. Thus, unlike radio and club sets, tonal design in film is not motivated *purely* by tonal continuity.

According to mid-twentieth-century Russian composer and music critic Leonid Sabaneev, early Hollywood film composers worked by an unofficial rule of thumb known as the "fifteen second rule," which held that successive film cues may be tonally unrelated as long as a minimum of fifteen seconds separates the cues.[8] After fifteen seconds of musical silence have passed, Sabaneev claimed, the listener has aurally forgotten the previous tonal center and is ready to receive a new tonal center without any disjunction. But obviously, what we have observed in

---

8. Sabaneev as quoted in Gorbman (1987: 90).

the films analyzed in this book far outstrip the confines of fifteen seconds, with tonal relationships spanning across hours. Even within the localized context of two *contiguous* same-key cues, tonal continuity is not necessarily the main goal (or even a factor), as we can see by revisiting the ending of *The Grand Budapest Hotel*. The music of the closing credits consists of three preexisting works ("S'Rothe-Zäuerli," "Kamarinskaya," and "Svetit Mesyats") all in the key of A major. One might assume that three A-major pieces were selected simply to maintain tonal continuity between contiguous cues, but notice that the middle cue, "Kamarinskaya," actually begins with a G-major introductory phrase. This introduction does lead to A major (through a rather circuitous path), but not before dislodging the sound of the previous cue's A-major tonality from our ears:

G: I–IV I–IV

   a: III–ii^{o7}–i–v^6–iv^6–V **A:** I

And even with regard to sound effects, there are countless instances where $\hat{1}/\hat{3}/\hat{5}$-pitched sound effects are deployed far more than fifteen seconds away from the nearest musical cue—observe, for example, the forty-five-second lapse between music and $\hat{1}$-$\hat{3}$ pitched sound effect in *On Golden Pond* (1981) shown in Video 36 (✶) and the forty-eight-second lapse in *When Harry Met Sally . . .* (1989) shown in Video 37 (✶). In the latter film, notice that the E♭-major song "It Had To Be You" ends unresolved on $\hat{2}$ as Harry (Billy Crystal) and Sally (Meg Ryan) squirm uncomfortably to sleep after their unexpected and ill-advised sexual encounter (a nice example of cadential frustration), and forty-eight seconds later, an ersatz resolution arrives in the form of a distant car horn honking on $\hat{1}$-$\hat{3}$, followed immediately by the $\hat{1}$-$\hat{3}$ of Marie's (Carrie Fisher) telephone ring as Sally calls her to discuss the messy situation. Clearly, filmic tonal design can go far beyond the simple logistical need for localized tonal continuity (via the "fifteen second rule"), providing long-range and *artistic* continuity and gestalt, as soundtracks continue to become more sophisticated as crafted entities.

FILMS THAT DON'T RESPOND WELL TO TONAL ANALYSIS

As with any analytical technique, some repertoire will provide better results than others. Although I have not found any particular film genre that consistently exhibits tonal design elements more than others, there *are* films that don't respond well to tonal analysis, as I demonstrate with a few examples. *American Graffiti* (1973) uses music (popular radio hits) as a central tool for establishing the milieu of early 1960s teen car culture, but its soundtrack makes scattershot use of virtually every key, with no salient emphasis on any one key and no noticeable key relationships.

The original scoring in *Atonement* (which we glanced at in chapter 5 for its use of sound effects) cycles quickly through numerous keys, making it difficult to define a "tonic." We could interpret this fleeting, fluctuating tonality as strongly suggestive of Briony's dangerously capricious nature, but aside from this generalization, it is difficult (at best, or artificial at worst) to discuss individual keys in this film. Mercurial modulation is the reason that films with the classic Hollywood wall-to-wall underscoring tend not to respond well to tonal analysis, because of the futility of declaring a definitive tonal center. On the other end of the spectrum, Morricone composed the entire soundtracks for *A Fistful of Dollars* and *For a Few Dollars More* in a single key (D minor), which certainly indicates tonal planning; but the sole use of a single key does not allow us to interpret key relationships either.

At the external (interwork) level, we can wonder why *this* particular key is employed (rather than, say, E minor) and speculate what intertextual allusions this specific key may evoke. But at the internal (intrawork) level, remaining exclusively in a single key saps that key of significance *within* the film, as we cannot locate any means (associativity, directionality, transposition, etc.) by which to interpret tonality in terms of the filmic narrative. The bottom line is that not every film will yield meaningful tonal interpretation, and it is important to resist the urge to force this—or any—methodology where it doesn't quite fit. Analytical judgment is always the key.

CLOSING THOUGHTS

Obviously there is much more that can be discussed about the music in these films beyond just key, but the purpose of this book is to illustrate how much we can gain by including key as one of the standard parameters for analyzing film music and sound.[9] When Theodor Adorno and Hanns Eisler dismissed the possibility of long-range tonality in film (in their 1947 monograph), their conception of tonality was based on the particular model of *tonal structure* that predominated (a narrow swath of) Western European instrumental art music. They posited "traditional tonality" and "atonality" as the two binary options for music (1947 [2005]: 123n2), but nowadays we understand tonality as a continuum that is continually evolving across musical eras and genres. (And, as I pointed out in chapter 1, the notion of a "traditional tonality" is itself problematic, since tonality has varied as much between the seventeenth and nineteenth centuries as it has between the nineteenth and twenty-first centuries.) What Adorno and Eisler found lacking (and didn't

---

9. For an example of how tonal analysis can augment and complement other methodologies, see Brian Jarvis (2023) for his nuanced exploration of the film *Barton Fink* (1991). Jarvis seamlessly incorporates associative tonality and cadential frustration into his analysis of the music's formal and thematic development with regards to the film's narrative/dramatic structure.

imagine was possible) in film music was "broad, well-planned harmonic canvases" that give a sense of "tonal organization" (1947 [2005]: 123n2)—and setting aside the specifics of their particular percept of tonality, I would argue that we have seen ample evidence of these characteristics in the film analyses we have explored throughout this book.

As to *how* an entire film soundtrack, with its many elements, could possibly turn out this way, I quote Charles Dickens who, as a postscript to *Our Mutual Friend*, describes "the relations of its finer threads to the whole pattern which is always before the eyes of the story weaver at his loom." Robert Bailey cites this Dickens quote in his discussion of Wagner's "preconceived plan of the entire work," in speculating how it was possible to tonally design something as massive as the seventeen-hour *Ring* cycle (1977: 61). Of course, a film soundtrack is much shorter than the *Ring*, but it involves dozens of people rather than one single Wagner at the helm. But then again, Wagner didn't have the software and digital tools for manipulating key/pitch at the touch of a button.

As to *why* a film soundtrack might turn out this way, I reiterate the closing thoughts of chapter 2 in pointing out that composers, arrangers, and the like often block chords at a keyboard to map out their musical plans, and that the tonal relationships common to Wagnerian (and European romantic) harmonic practice continue to inform and influence Hollywood film music. Whether filmmakers are consciously making decisions about keys or unconsciously responding to harmonic relationships that just "feel right" (to reiterate Joel Coen), the privilege of interpreting film tonality belongs to the viewer-listener.[10] The goal is not to reveal The One True Interpretation—musical analysis is unique and intriguing specifically because it "allow[s] multiple potential meanings and demand[s] none in particular" (Abbate 2004: 534). The point of this kind of analysis is simply to create an "inter-subjectively plausible and revealing" interpretation (Lehman 2012b: 104) that will "convince the reader to come along for the ride" (Latham 1997: 17).

An interpretation of film tonality is a primarily intellectual (rather than sensory) undertaking, since most of us do not have the ability to *physically hear* tonal relationships across large spans of time. But as with a study of J. S. Bach's B-A-C-H cryptogram in his *Art of Fugue*, we don't perceive every iteration organically, but we learn to appreciate them through analysis. And even with those B-A-C-H iterations that we cannot *hear* because they whiz by too quickly, or are buried too deeply within the texture, or are transformed too covertly, we take satisfaction in *knowing* they are there. In the same vein, we might not *hear* the directional uplift from the opening E♭ minor to the closing E minor in *The Graduate*, symbolizing Benjamin's uprising, but once we *know* about it, it's hard not to imagine we can hear and feel that directional tonality. Iterative, investigative analysis greatly

---

10. As Orgeron puts it, "texts are authored, finally, by the reader" (2007: 59).

increases our engagement with a work and satisfies our love of "the cryptographic sublime," humans' insatiable desire to uncover narrative meaning and "reveal something inaccessible" (Abbate 2004: 524 and 520).

Claudia Gorbman unleashed the analytical floodgates in 1987 when she drew our attention to the "unheard" nature of film music. Twenty years later, she made the pronouncement that "melodies are no longer unheard," because the "strictures and underlying aesthetic" of the image-narrative-music relationship have changed such that film music is now solidly within our notice (Gorbman 2007: 151). But what *has* remained unnoticed thus far in film music is key, and its role in contributing to the filmic narrative and amalgamating the soundtrack into a cohesive entity. To reiterate my points from the end of chapter 1, the essence of film tonality boils down to this: Does key matter? Do we gain anything from considering the tonal layout of a film? Would we lose anything if a particular cue was displaced to a different key? Would we lose anything if all cues in all films were restricted to the key of C major/minor?

Of course, we certainly *can* continue to navigate films without considering their key constellations—but to adapt the incomparable Carl Sagan, it seems like an awful waste of keys.[11]

---

11. The saying ubiquitously associated with Carl Sagan—"If it is just us, it seems like an awful waste of space"—comes from the 1997 film adaptation of his 1985 book *Contact*. Sagan espoused this basic sentiment throughout his illustrious career, and he originally adapted it from Thomas Carlyle, during the "Life Beyond Earth and the Mind of Man" symposium held at Boston University on November 20, 1972: "Thomas Carlyle, a somewhat crusty old fellow, upon thinking about the stars, said: 'If they be inhabited, what a scope for misery and folly. If they be not inhabited, what a waste of space'" (Sagan 1973: 5).

APPENDIX

Working Method for Creating a "Tonal Score"

To analyze a Schubert song, one usually begins by opening a copy of the score. A film soundtrack doesn't have a score—even if one could access the manuscripts for the original underscoring, that wouldn't include the preexisting music and sound effects in the film. So to tonally analyze a film, we first begin by taking inventory of all the keys and pitches in the soundtrack, constructing what I call the "tonal score."

The basic tools for constructing a tonal score include a digital copy of the film, a computer (to allow for frame-by-frame analysis and image capture), headphones (to hear the subtle details in a soundtrack that evade bare ears), a digital or online keyboard (to ensure that pitch is always judged against a consistent standard—e.g., A440), and a spreadsheet (for taking the detailed notes that will become the notational trace of the film's soundtrack).[1] You want to construct the most complete tonal score possible, because you don't know beforehand which elements might be important for your analysis—so every musical and pitched event should be notated in detail. I use an ordinary Microsoft Excel spreadsheet for my tonal scores, but any spreadsheet software will do. The columns I include in my spreadsheet are:

- **START TIME** (HH:MM:SS format, and with the text formatted to the "Time" category)

---

1. Also helpful, though definitely not required, would be pitch-shifting software (such as Melodyne or Autotune) that allows extraction of pitch content as well as a digital audio workstation (DAW) or other software that allows the application of frequency filters to isolate sounds. Transcription software (such as Transcribe!) could also be helpful in allowing the user to identify pitches as well as slow down passages without changing pitch. (I myself did not have access to this kind of software during my own work, but it certainly would have been useful!)

- **KEY/PITCH** (e.g., A major for keys, E-G* for pitches; I include the asterisk to make the non-keys easily visually distinguishable from keys)[2]
- **NAME** (composer/artist and cue title for music; sound source for sound effects—e.g., car horn, phone ring, etc.)
- **TYPE** (original, preexisting, sound effect)
- **NARRATIVE CONTEXT** (description of the concurrent narrative events)
- **NOTES** (any questions, hunches, observations you make while you're gathering data and might wish to investigate later)

There are two optional columns you can add to your spreadsheet (directly after START TIME), if you decide to create a "tonal graph" or "tonal staff" (discussed in the "Logistics" section of chapter 1):

- **END TIME** (HH:MM:SS format, and with the text formatted to the "Time" category)
- **DURATION** (auto-generated by populating the column with a subtraction formula between the START TIME and END TIME column values; make sure the text in this column is also formatted to the "Time" category)

Tracking end-times for every cue and sound effect is a time-consuming process, so these columns can be omitted (or left unpopulated) if you don't plan to create graphs or staffs.

Set an AutoFilter on each column so that you can instantly sort the spreadsheet data according to the different parameters. For example, you may sort the spreadsheet by the KEY column, to see if any key is used far more often than others. You might then AutoFilter to show *only* the cues in that particular key, to ascertain whether that key consists of multiple iterations of the same few musical works or a diverse collection of different works. You may sort by TYPE, to explore the relationship between preexisting and original works. This exploration process is particularly helpful for isolating and interpreting sound effects, which can be difficult to make sense of when they are scattered across an entire soundtrack. Synthesizing the data in this way lets you explore it from different perspectives, which allows trends and patterns to surface from a sea of unsorted data.

The NOTES column is where you should write any early observations, impressions, and questions that occur to you during the data-gathering phase. For instance, you might start to sense the associative qualities of a key, or a possibly meaningful relationship between two keys, or you might notate the scale degrees potentially being represented by sound effects. By the time you're done going through the entire film, those early inchoate impressions will be lost in a wash of other information, so it's better to jot something down with a question

---

2. Enharmonic distinctions are not of vital importance for our purposes, so sharp/flat designations can be made to fit the context. Example #1: if a hunting horn sounds an E♭/D♯ pitch, it would make sense to designate this as an E♭ because of this instrument's given properties. Example #2: if a B-major cue is followed by a cue in E♭ minor/D♯ minor, it might be helpful to designate this key as D♯ minor to show the relationship between the two keys.

mark in the NOTES column than to forget about it and have to rewatch the film to recall it. You will want to consult closely with the NARRATIVE CONTEXT column, to trace connections between harmonic events and dramatic events, and NOTES is where you'll make note of these connections. The NOTES column is really where your analysis starts to take shape.

In the types of films best suited to tonal analysis, most cues are unmistakably in a particular key and straightforward to tonally assess.[3] But for cues that do present a challenge, there are several different ways to diagnose tonality. Musical fragments may be examined for common tonic-seeking components such as $\hat{7}-\hat{1}$ or $\hat{5}-\hat{1}$ melodic motion or dominant–tonic harmonic motion. If a cue is excerpted from a section of extended tonicization or modulation (i.e., the tonic of the excerpt included in the film is different than the global tonic of the overall work), the temporary tonic of the excerpt should be privileged, since that is what we hear in the filmic context. If a musical excerpt is too short to adequately develop a sense of harmonic inclination, and there are no dominant–tonic implications, we can make our assessment based on the ear's natural inclination to favor the first pitch/triad as tonic.[4]

A cue containing a modulatory passage within it can either be (1) designated by the starting/ending key (if the entire passage is tonally closed), (2) designated by its separate starting and ending points (if the passage is tonally open), or (3) parsed into individual key areas (if there are distinct, self-sufficient modulatory units). In works from post-tonal genres or non-Western musical traditions, the sense of tonic can sometimes be estimated via *tonality by assertion*, which draws upon such factors as repetition, voicing, register, metric placement, rhythmic duration, dynamics, "melodic goal-directedness" (Everett 2004), scalar collections, hierarchical relationships, and overtones, among others.[5] If a musical selection simply defies any sort of tonal classification, the best course of action is to note it as such and avoid the temptation to force a tonal label on it. (However, I have found this scenario to be rare in mainstream Hollywood films.)

Since musical works (both original and preexisting) are frequently transposed to different keys, it is best not to assume that any musical piece is in its original key—even if the same piece is repeated multiple times. (And the same goes for sound effects.) In the case of preexisting works, you can consult the original recording or the score to cross-check the work's original key. If the work is *not* in its original key, you have an interesting matter to pursue: Was it transposed expressly for the film, and if so, why? First you should consult recordings

---

3. Films featuring mostly tonal music work best, especially if the music is broken up into distinct cues rather than continuous wall-to-wall underscoring.

4. David Huron has found in laboratory experiments that "listeners were most easily able to imagine an isolated tone as the tonic" (2006: 65). And almost two hundred years earlier, Gottfried Weber stated that "it is natural that, in the beginning of a piece of music, when the ear is as yet unpreoccupied with any key, it should be inclined to assume as the tonic any major or minor three-fold harmony that first presents itself" (1817 [1851]: 333).

5. The earliest use of the phrase "tonality by assertion" appears in Eric Salzman's book, *Twentieth-Century Music* (1974: 29). For further discussions of alternative methods for diagnosing tonality, works by Daniel Harrison (2016 and 1994) and Walter Everett (2004) are excellent starting points.

of the work to determine if the work was transposed to that key for the film or not.[6] For example, the Sarabande from Bach's French Suite No. 3 is written in the key of B minor but appears in *Persuasion* (1995) in the key of C♯ minor. No other recordings or versions of this work occur in C♯ minor, so this provokes us to interpret *why* the key has been altered (which we did in chapter 3).

Identifying preexisting music can often present a challenge, as soundtrack listings (in the closing credits or on IMDB, or in the commercial soundtrack album) are sometimes inaccurate and *almost always* incomplete. Sometimes the artist and recording used in the *film* are different from what appears in the *album*—and many works are left out altogether. And not infrequently, certain works listed in the soundtrack are nowhere to be found in the film. Since soundtrack listings are of so little help, one must rely on a variety of alternative methods for determining unidentified musical works used in a film. If there are lyrics involved, the matter can usually be resolved by using a Google search for a specific line of the lyrics (enclosed in quotes). Without lyrics, it becomes more difficult. Quite often, one is reduced to deducing the identity of a classical work, for example, by guessing at the composer, instrumentation or ensemble, movement type, combining this information with the apparent (audible) key, and then searching through the works of that surmised composer to see if there is a match (e.g., searching through Mozart's string quartets for an allegro movement in B♭ major). The process of cue-sleuthing can be aided by sources such as music-identification applications (e.g., Shazam), soundtrack-focused websites (e.g., What-Song), online playlists (e.g., YouTube), web forums, blog posts, and even closed captions and screenplays (in which specific musical works are sometimes identified by name). But in the end, you can still identify the key of the music even if you can't identify the name/composer, so don't let that stop you.

Even the identification of original works (composed expressly for the film) can present a challenge. When original works *are* listed in the soundtrack album (which is not always guaranteed), one must usually parse beyond the given name designations, because several distinct themes are lumped together into one continuous track.[7] And often, one piece of distinct thematic material may appear in several differently named tracks, sometimes with different voicing or instrumentation. These factors complicate the taxonomy of film themes and complicate the tonal score construction process. So you may find it helpful to classify themes as discrete entities and assign them separate names if they are used discretely in the film context, regardless of soundtrack designations.

With sound effects, we note their pitches rather than keys. (A sound effect only receives a *key* designation if it contains multiple pitches that are organized in a way that exhibits clear scalar or chordal organization; for example, church bells cycling through a descending C-major scale could be labeled as C major, or a car horn featuring the pitches G, B, and D

---

6. Sometimes, for instance, an old historic recording might sound a half step lower than the work's actual key, so you should take this into account, rather than assuming it was *transposed* a half step lower expressly for the film.

7. It makes for a more polished soundtrack album to weave themes into a cohesive, self-contained medley rather than include a handful of disconnected twenty-second snippets of instrumental themes—especially since not every theme has a clearly delineated beginning and ending, making for an awkward standalone track.

could be labeled as G major.) The main action is to examine how the pitches of sound effects interact with and relate to the keys of the music around them. For example, in a film that associatively features the key of E minor to signify danger, it might be noteworthy that a grandfather clock that strikes an E when a suspicious character enters the room. In such a case, that E-pitched sound effect functions as $\hat{1}$. If a B-pitched sound effect was consistently used in that film, it could be functioning as $\hat{5}$—or perhaps a persistent D♯ pitch might be serving as $\hat{7}$. The important thing is to make sure that a pitch has a palpable role in the narrative, before assigning it a functional label or designating it as part of an associative system. And of course, even without an overarching framework such as associative tonality, sound effects can be analyzed simply in terms of how their pitches interact with one another and with surrounding music.

The result of the data-gathering process (which takes several hours because of all the stop-and-start, identification, and documentation) is a spreadsheet that serves as a complete record of all pitched elements of a soundtrack. This tonal score can now be analyzed for patterns and anomalies. As discussed, not every film will manifest elements of systemic tonal organization, but as with any other analytical endeavor, you won't know until you dig in!

GLOSSARY

absolute music	instrumental music unaccompanied by any overt literary, pictorial, or dramatic references. *See also* program music.
acousmatic	a sound whose source is not visually revealed.
Aeolian cadence	the conclusion of a musical phrase, prevalent in pop and rock music, consisting of ♭VI—♭VII—I(or i).
anempathetic sound	the sound that is markedly indifferent or even contradictory to the visual imagery.
appellative	the strong, resolution-seeking energy created by combining $\hat{4}$ and $\hat{7}$, usually in the context of a dominant-seventh chord, in which these scale degrees form the chordal seventh and third, respectively. In Western tonality this unstable tritone strongly "wants" to resolve with $\hat{4} \rightarrow \hat{3}$ and $\hat{7} \rightarrow \hat{1}$.
associative tonality	specific keys or chords are consistently associated with specific narrative elements, creating a system of musical-narrative symbolism.
auteur	the director considered to be the author of a film, due to the strong control they exert over every aspect of the filmmaking process, creating films that clearly manifest their artistic signature and individual style.
authentic cadence	the conclusion of a musical phrase with dominant–tonic harmonic motion (V–I or i). Provides the strongest sense of closure in Western music. The term "perfect" applies to an authentic cadence if the top voice resolves stepwise to $\hat{1}$ and the bottom voice resolves from $\hat{5} \rightarrow \hat{1}$

background level	the deepest layer of a musical work in which only the most important structural elements are represented. *See also* foreground level, middleground level, and Schenkerian analysis.
barn-door wipe	a shot transition in which the shot either slides open or slides shut.
blue notes	the flatted versions of diatonic scale degrees, commonly 3, 5, and 7—especially when used in conjunction with the diatonic versions.
coherence	the tight construction of a work such that clear connections can be drawn between the parts and the whole. *See also* organicism.
compilation score	a soundtrack consisting almost entirely of preexisting music.
cue	a musical selection in a film.
development section	the middle section of a work (in sonata form) characterized by instability and activity as the main thematic material of the work (from the exposition) undergoes restless rumination and reworking, before finally reestablishing stability with the return of the main thematic material (the recapitulation).
diegetic	sound that originates in the narrative world and can thus be heard by the characters in it. *See also* nondiegetic and metadiegetic.
directional tonality	a musical work ends in a different key than that in which it began.
dissolve	the overlapping transition between shots so that one fades out as the other fades in.
double tonality	the interplay between two tonics that coexist as equals.
drastic	a perception generated by immediate, physical sensory discernment. *See also* gnostic.
esthesic	a perspective that privileges the perception/interpretation of the reader. *See also* poietic.
fantastical gap	the ambiguous space between diegetic and nondiegetic sound/music.
foreground level	the surface layer of a musical work in which a great amount of detail is included. *See also* background level, middleground level, and Schenkerian analysis.
functional tonality	a musical system in which the chords and pitches within a key are arranged in hierarchical relationships to the tonic (pre-dominant, dominant, and tonic functions [P, D, and T]). Musical motion is governed by this hierarchy, which

	consists of relative senses of stability and instability with respect to the tonic.
gnostic	a perception that is mediated by the thought process and intellectual consideration. *See also* drastic.
half cadence	the conclusion of musical phrase with a dominant chord (V), creating an open-ended feeling that implies further activity in search of the more conclusive resolution of an authentic cadence.
hexatonic pole	the harmonic relationship between a major and minor triad, where the minor triad's root is located four semitones (a major third) below the root of the major triad (e.g., C major and A♭ minor). These two triads share no pitches in common with one another, and together their six pitches form a hexatonic collection (also known as the "augmented scale"). In Western music the hexatonic pole relationship is often associated with the Uncanny and dark mysterious implications.
hierarchy	*see* functional tonality.
intertextuality	the relationship between two (or more) different works, in which one work references another.
intertitle	a placard of written text inserted between scenes in a film. Also known as a title card.
jump cut	an editing technique that conveys the passage of time by jumping between different shots of the same setting.
leitmotif	a recurrent musical idea that acquires symbolic meaning through its association with a particular character/object/location, and evokes that associative meaning every time it recurs.
Leittonwechsel	a transformation between two triads in which the root of a major triad is lowered by a half-step (or the fifth of a minor triad is raised by a half-step). Thus a C-major triad transforms into an E-minor triad (or a C-minor triad transforms into an A♭-major triad). *See also* neo-Riemannian, transformation.
lydian	a mode similar to the major mode scale but with raised $\hat{4}$.
metadiegetic	sound that originates in the mind or memory of a character and thus cannot be heard by other characters in the scene. *See also* diegetic and nondiegetic.
metric displacement	musical meter consists of two or more levels of pulse that align to form a hierarchy of "strong" and "weak" beats. Metric displacement occurs when the "main" beat (or tactus) of one layer shifts forward or backward, so that two

	metric layers are now misaligned (out of phase) with one another.
mickey-mousing	music written and synchronized to closely mimic onscreen action. So named because of its prevalent use in early Disney cartoons.
middleground level	the intermediary layer of a musical work, in which selected important details adorn the structural elements. *See also* background level, foreground level, and Schenkerian analysis.
mise-en-bande	the sonic counterpart to *mise-en-scène*, encapsulating the entire auditory fabric of a film.
mise-en-scène	the visual design of a shot, including elements such as scenery, props, costumes, lighting, framing, and staging.
Mixolydian	a mode similar to the major mode scale but with lowered $\hat{7}$.
monotonality	one tonic controls an entire musical work, so that any deviations (e.g., modulations) from that tonic are in service of prolonging the tonic. *See also* prolongation.
montage	a sequence that uses music (usually nondiegetic) to weave together separate shots into a continuous whole.
neo-Riemannian	an analytical perspective that focuses on the relationship and motion between chords, rather than focusing on each chord's relation to a tonic. This approach is often applied to highly chromatic music of the late nineteenth century, in which diatonic functional logic is supplemented (and at times supplanted) by pan-triadic chromaticism. *See also* transformation.
nondiegetic	sound that exists outside the narrative world and can thus be heard only by the film viewers (not the film characters). *See also* diegetic and metadiegetic.
organicism	the dynamic balance between the parts and the whole, between unity and diversity. From this perspective, everything in a work germinates from a core kernel such that all the diverse elements of the work may be traced back to this kernel.
original music	music that was written specifically for a film. *See also* preexisting music.
parallel interrupted period	a musical phrase structure that divides into two halves, each beginning with similar thematic material. The first half ends with a half cadence, and the second half ends with a perfect authentic cadence.

paratext	the liminal elements surrounding the main body of a work (e.g., book or film), forming the interface between work, author, publisher, and reader. For example, in a book, paratext includes its cover, foreword, index, and so forth.
Picardy third	a major-mode tonic triad ending a piece written in minor mode.
plagal	the harmonic progression of a subdominant chord to a tonic chord (IV–I or any combination of the major and minor versions of these chords).
poietic	the perspective that privileges what the writer created/intended. *See also* esthesic.
preexisting music	music used in a film that was written independently of the film. For example, a Beatles song or a folk tune or a Beethoven symphony that is used in a film soundtrack. *See also* original music.
program music	instrumental music that depicts a narrative or imagery, usually through means of an expressive title or accompanying text (known as the "program"). *See also* absolute music.
prolongation	one chord/pitch is extended over time, so that deviations from that chord/pitch can be interpreted as embellishments of that fundamental chord/pitch. *See also* monotonality and Schenkerian analysis.
promissory note	a musical element (e.g., pitch or chord) that draws attention to itself in its context, with the expectation that it will achieve later resolution.
Schenkerian analysis	an analytical method that distills the fundamental structure of a work by stripping away all non-essential, non-structural elements of the music. This approach is based on the notion that compositions are fundamentally monotonal, and that an entire musical work can be demonstrated to be an embellished prolongation of a fundamental I–V–I harmonic progression.
screen aspect ratio	the relationship between the width and height of a film image.
sound advance	a type of sound bridge in which the sound for a scene begins before the visual image of the scene begins. *See also* sound lag.
sound bridge	a sound editing technique in which sounds are used to smooth the visual discontinuity of a scene transition.
sound lag	a type of sound bridge in which the sound from one scene continues to linger after the visual transition to the next scene. *See also* sound advance.

Sturm und Drang	translating to "storm and stress" (based on a late eighteenth-century German artistic movement), the expression of emotional turbulence and unrest.
tonal design	the deployment of keys throughout a work. *See also* tonal structure.
tonality by assertion	the tonal center established not by functional harmony, but by factors such as repetition, voicing, register, metric placement, rhythmic duration, dynamics, melodic goal-orientation, scalar collections, hierarchical relationships, and overtones, among others.
tonally closed	a work/passage that begins and ends in the same key. *See also* tonally open.
tonally open	a work/passage that begins and ends in different keys. *See also* tonally closed.
tonal structure	the deployment of keys throughout a work, under the hierarchical influence of a single governing tonic. *See also* monotonality, prolongation, and tonal design.
transformation	an operation through which one chord transforms into another, rather than one static object being replaced by a separate static object. For example, a C-major triad transforms into an A♭-major triad by shifting E→E♭ and G→A♭. *See also* neo-Riemannian.
triadic chromaticism	a style of composition popularized by 1990s Hollywood action films in which major/minor triads are used to craft chromatic harmonic progressions (i.e., progressions that do not abide by diatonic and functional logic). *See also* transformation and Neo-Riemannian.
twelve-tone music	a compositional method (developed during the atonal movement of the twentieth century) that uses all twelve tones of the chromatic scale in ordered series. Series are used, reused, and manipulated in different ways over the course of a composition, granting coherence to the overall work. Triads and functional diatonic harmony are not requisite (or often even present) in twelve-tone music. (Also sometimes referred to as serialism, of which it is a subtype.)
underscoring	nondiegetic music played beneath dialogue.

BIBLIOGRAPHY

Abbate, Carolyn. 2004. "Music—Drastic or Gnostic?" *Critical Inquiry* 30(3): 505–536.
Abbate, Carolyn, and Roger Parker. 1989. *Analyzing Opera: Verdi and Wagner*. Berkeley: University of California Press.
Adorno, Theodor, and Hanns Eisler. 1947 (2005). *Composing for the Films*. London: Continuum.
Agawu, Kofi. 2009. *Music as Discourse: Semiotic Adventures in Romantic Music*. New York: Oxford University Press.
Altman, Rick, McGraw Jones, and Sonia Tatroe. 2000. "Inventing the Cinema Soundtrack: Hollywood's Multiplane Sound System." In *Music and Cinema*, edited by James Buhler, Caryl Flinn, and David Neumeyer, 339–359. Middletown, CT: Wesleyan University Press.
Bailey, Robert. 1969. "The Genesis of Tristan und Isolde and a Study of the Sketches and Drafts for the First Act." PhD dissertation, Princeton University.
———. 1977. "The Structure of the *Ring* and Its Evolution." *19th-Century Music* 1(1): 48–61.
———. 1985. "An Analytical Study of the Sketches and Drafts." In *Prelude and Transfiguration from Tristan and Isolde*, edited by Robert Bailey, 113–146. New York: W. W. Norton.
BaileyShea, Matt. 2007. "The Hexatonic and the Double Tonic: Wolf's 'Christmas Rose.'" *Journal of Music Theory* 51(2): 187–210.
Balthazar, Scott. 1996. "Plot and Tonal Design as Compositional Constraints in *Il Trovatore*." *Current Musicology* 60-61: 51–78.
Barthes, Roland. 1977. "The Death of the Author." In *Image-Music-Text*, translated by Stephen Heath, 142–148. London: Fontana.
Beach, David. 1993. "Schubert's Experiments with Sonata Form: Formal-Tonal Design versus Underlying Structure." *Music Theory Spectrum* 15(1): 1–18.
Bernard, Laure (producer), and Rani Khanna (director). 2007. *Music by Gabriel Yared* [Film]. Kultur Video.

Bloom, Harold. 1973. *The Anxiety of Influence: A Theory of Poetry*. London: Oxford University Press.
Bordwell, David, and Kristin Thompson. 1997. *Film Art: An Introduction*. New York: McGraw-Hill Companies, Inc.
Bribitzer-Stull, Matthew. 2001. "Thematic Development and Dramatic Association in Wagner's 'Der Ring des Nibelungen.'" PhD dissertation, University of Rochester.
———. 2006a. "The A♭–C–E Complex: The Origin and Function of Chromatic Major Third Collections in Nineteenth-Century Music." *Music Theory Spectrum* 28(2): 167–190.
———. 2006b. "The End of Die Feen and Wagner's Beginnings: Multiple Approaches to an Early Example of Double-Tonic Complex, Associative Theme and Wagnerian Form." *Music Analysis* 25(3): 315–340.
———. 2012. "From Nibelheim to Hollywood: The Associativity of Harmonic Progression." In *The Legacy of Richard Wagner*, edited by Luca Sala, 157–183. Turnhout: Brepols.
———. 2015. *Understanding the Leitmotif: From Wagner to Hollywood Film Music*. Cambridge: Cambridge University Press.
Brown, Matthew. 1989. "Isolde's Narrative: From Hauptmotiv to Tonal Model." In *Analyzing Opera: Verdi and Wagner*, edited by Carolyn Abbate and Roger Parker, 180–201. Berkeley: University of California Press.
Brown, Royal. 1982. "Herrmann, Hitchcock, and the Music of the Irrational." *Cinema Journal* 21(2): 14–49.
———. 1994. *Overtones and Undertones: Reading Film Music*. Berkeley: University of California Press.
Buhler, James. 2001. "Analytical and Interpretive Approaches to Film Music (II): Analyzing Interactions of Music and Film." In *Film Music: Critical Approaches*, edited by Kevin Donnelly, 39–61. Edinburgh: Edinburgh University Press.
———. 2010. *Hearing the Movies: Music and Sound in Film History*. New York: Oxford University Press.
———. 2019. *Theories of the Soundtrack*. New York: Oxford University Press.
Burton, David. 2011. *The History of Mathematics: An Introduction*, 7th edition. New York: McGraw-Hill.
Caplin, William E. 1998. *Classical Form: A Theory of Formal Functions for the Instrumental Music of Haydn, Mozart, and Beethoven*. Oxford: Oxford University Press.
Carlton, Dave. 2012, June 6. "I analyzed the chords of 1300 popular songs for patterns. This is what I found." *The Hooktheory Blog*. www.hooktheory.com/blog/i-analyzed-the-chords-of-1300-popular-songs-for-patterns-this-is-what-i-found/.
Casanelles, Sergi. 2016. "Mixing as a Hyperorchestration Tool." In *The Palgrave Handbook of Sound Design and Music in Screen Media*, edited by Liz Greene and Danijela Kulezic-Wilson, 55–72. London: Palgrave Macmillan.
Chion, Michel. 1994. *Audio-Vision: Sound on Screen*, translated by Claudia Gorbman. New York: Columbia University Press.
Cochran, Alfred. 1986. "Style, Structure, and Tonal Organization in the Early Film Scores of Aaron Copland." PhD dissertation, Catholic University of America.
———. 1990. "The Spear of Cephalus: Observations on Film Music Analysis." *Indiana Theory Review* 11: 65–80.

Coen, Joel, and Ethan Coen. 2006. *The Coen Brothers: Interviews*, edited by William Rodney Allen. Jackson: University Press of Mississippi.

Cohn, Richard. 2004. "Uncanny Resemblances: Tonal Signification in the Freudian Age." *Journal of the American Musicological Society* 57(2): 285–323.

———. 2006. "Hexatonic Poles and the Uncanny in Parsifal." *Opera Quarterly* 22(2): 230–248.

———. 2012. *Audacious Euphony*. New York: Oxford University Press.

Coleman, Lindsay. 2017. "Dario Marianelli Interview." In *Contemporary Film Music: Investigating Cinema Narratives and Composition*, edited by Lindsay Coleman and Joakim Tillman, 139–153. London: Palgrave Macmillan.

Collis, Clark. 2017, December 4. "How Director Edgar Wright Steered *Baby Driver* to Global Success." *Entertainment Weekly*. https://ew.com/movies/2017/12/04/edgar-wright-baby-driver-success-sequel-spacey/.

Cone, Edward T. 1982. "Schubert's Promissory Note: An Exercise in Musical Hermeneutics." *19th-Century Music* 5(3): 233–241.

Cook, Nicholas. 1987. "The Perception of Large-Scale Tonal Closure." *Music Perception: An Interdisciplinary Journal* 5(2): 197–205.

Cooper, Michael. 2017, February 21. "Hear How 'Moonlight' Got Its Sound: Violins, Chopped and Screwed." *New York Times*. www.nytimes.com/2017/02/21/arts/music/moonlight-movie-score-music-oscars.html.

Decker, Todd. 2012. "The Musical Mr. Ripley." *Music, Sound, and the Moving Image* 6(2): 185–207.

Doll, Christopher. 2011. "Rockin' Out: Expressive Modulation in Verse-Chorus Form." *Music Theory Online* 17(3). https://mtosmt.org/issues/mto.11.17.3/mto.11.17.3.doll.html.

———. 2017. *Hearing Harmony: Toward a Tonal Theory for the Rock Era*. Ann Arbor: University of Michigan Press.

Drabkin, William. 1982. "Characters, Key Relations and Tonal Structure in Il Trovatore." *Music Analysis* 1(2): 143–154.

Everett, Walter. 2004. "Making Sense of Rock's Tonal Systems." *Music Theory Online* 10(4). https://mtosmt.org/issues/mto.04.10.4/mto.04.10.4.w_everett.html

Fagen, Donald. 1989. "Ennio Morricone? Ah, Bellissimo!" *Premiere* (August 1989): 106–107.

Falsetto, Mario. 2013. *Anthony Minghella: Interviews*. Jackson: University Press of Mississippi.

Fish, Stanley. 1980. *Is There a Text in This Class? The Authority of Interpretive Communities*. Cambridge: Harvard University Press.

Foucault, Michel. 1979. "What Is an Author?" In *Textual Strategies*, edited by Josue V. Harari, 141–160. Ithaca, NY: Cornell University Press.

Gauldin, Robert. 2015. "Associative Transposition in Wagner's Ring." Paper presented at the annual conference of the Music Theory Society of New York State (MTSNYS), Binghamton, NY, April 11.

Gjerdingen, Robert. 1999. "An Experimental Music Theory?" In *Rethinking Music*, edited by Nicholas Cook and Mark Everist, 161–170. Oxford: Oxford University Press.

Gorbman, Claudia. 1987. *Unheard Melodies: Narrative Film Music*. Bloomington: Indiana University Press.

———. 2007. "Auteur Music." In *Beyond the Soundtrack: Representing Music in Cinema*, edited by Daniel Goldmark, Lawrence Kramer, and Richard Leppert, 149–162. Berkeley: University of California Press.

Guck, Marion A. 2006. "Analysis as Interpretation: Interaction, Intentionality, Invention." *Spectrum* 28: 191–209.

Hall, Jacob. 2017, June 28. "*Baby Driver*: The Edgar Wright Interview." */Film*. www.slashfilm.com/baby-driver-edgar-wright-interview/.

Harrison, Daniel. 1994. *Harmonic Function in Chromatic Music: A Renewed Dualist Theory and an Account of Its Precedents*. Chicago: University of Chicago Press.

———. 2016. *Pieces of Tradition: An Analysis of Contemporary Tonal Music*. New York: Oxford University Press.

Hartley Andrew James. 2009. "Character, Agency and the Familiar Actor." In *Shakespeare and Character: Palgrave Shakespeare Studies*, edited by Paul Yachnin and Jessica Slights, 158–176. London: Palgrave Macmillan.

Hatten, Robert. 1991. "On Narrativity in Music: Expressive Genres and Levels of Discourse in Beethoven." *Indiana Theory Review* 12: 75–98.

———. 1994. *Musical Meaning in Beethoven: Markedness, Correlation, and Interpretation*. Bloomington: Indiana University Press.

Heldt, Guido. 2013. *Music and Levels of Narration in Film: Steps across the Border*. Bristol, UK: Intellect.

Hepokoski, James. 1989. "Verdi's Composition of Otello: The Act II Quartet." In *Analyzing Opera: Verdi and Wagner*, edited by Carolyn Abbate and Roger Parker, 125–149. Berkeley: University of California Press.

Hepokoski, James, and Warren Darcy. 2006. *Elements of Sonata Theory: Norms, Types, and Deformations in the Late-Eighteenth-Century Sonata*. Oxford: Oxford University Press.

Holbrook, Morris. 2005. "The Ambi-Diegesis of 'My Funny Valentine.'" In *Pop Fiction: The Song in Cinema*, edited by Steve Lannin and Matthew Caley, 47–62. Bristol: Intellect Books.

Hubbert, Julie. 2014. "The Compilation Soundtrack from the 1960s to the Present." In *The Oxford Handbook of Film Music Studies*, edited by David Neumeyer, 291–318. New York: Oxford University Press.

Huron, David. 2006. *Sweet Anticipation: Music and the Psychology of Expectation*. Cambridge: MIT Press.

Iser, Wolfgang. 1978. *The Act of Reading: A Theory of Aesthetic Response*. Baltimore, MD: Johns Hopkins University Press.

Jarvis, Brian Edward. 2023. "Prioritizing Narrative Structure in Large-Scale Film-Music Analysis: A Case Study of Dramatic Irony in *Barton Fink*." *Music Theory Online* 29(1). https://mtosmt.org/issues/mto.23.29.1/mto.23.29.1.jarvis.html.

Kashner, Sam. 2008, February 25. "Here's to You, Mr. Nichols: The Making of *The Graduate*." *Vanity Fair*. www.vanityfair.com/news/2008/03/graduate200803.

Katz, Adele. 1945. *Challenge to Musical Tradition: A New Concept of Tonality*. New York: Alfred A. Knopf.

Keller, Hans. 2006. *Film Music and Beyond: Writings on Music and the Screen, 1946–59*, edited by Christopher Wintle. London: Plumbago Books and Arts.

Kinderman, William. 1980. "Dramatic Recapitulation in Wagner's Götterdämmerung." *19th Century Music* 4(2): 101–112.

Klein, Michael L. 2005. *Intertextuality in Western Art Music*. Bloomington: Indiana University Press.

Kmet, Nicholas. 2018. "Remote Control: Collaborative Scoring and the Question of Authorship." *Revue musicale OICRM* 5(2): 1–14.

Korsyn, Kevin. 1991. "Towards a New Poetics of Musical Influence." *Music Analysis* 10(1/2): 3–72.

Krebs, Harald. 1981. "Alternatives to Monotonality in Early Nineteenth-Century Music." *Journal of Music Theory* 25(1): 1–16.

———. 1996. "Some Early Examples of Tonal Pairing: Schubert's 'Meeres Stille' and 'Der Wanderer.'" In *The Second Practice of Nineteenth-Century Tonality*, edited by William Kinderman and Harald Krebs, 17–33. Lincoln: University of Nebraska Press.

Kubernik, Harvey. 2006. *Hollywood Shack Job: Rock Music in Film and on Your Screen*. Albuquerque: University of New Mexico Press.

Kulezic-Wilson, Danijela. 2016. "Musically Conceived Sound Design, Musicalization of Speech and the Breakdown of Film Soundtrack Hierarchy." In *The Palgrave Handbook of Sound Design and Music in Screen Media*, edited by Liz Greene and Danijela Kulezic-Wilson, 429–444. London: Palgrave Macmillan.

———. 2017. "Sound Design and its Interactions with Music: Changing Historical Perspectives." In *The Routledge Companion to Screen Music and Sound*, edited by Miguel Mera, Ronald Sadoff, and Ben Winters, 127–138. New York: Routledge Press.

———. 2020. *Sound Design Is the New Score*. Oxford: Oxford University Press.

Laing, Heather. 2003. *Gabriel Yared's The English Patient: A Film Score Guide*. Lanham, MD: Scarecrow Press.

———. 2007. *The Gendered Score: Music and Gender in 1940s Melodrama and the Woman's Film*. Aldershot: Ashgate Publishing Limited.

Latham, Edward. 1997. "Review of Ethan Haimo's Article, 'Atonality, Analysis, and the Intentional Fallacy,' *Music Theory Spectrum* 18.2 (Fall 1996), 167–199." *Music Theory Online* 3(2). https://mtosmt.org/issues/mto.97.3.2/mto.97.3.2.latham.html.

———. 2008. *Tonality As Drama: Closure and Interruption in Four Twentieth-Century American Operas*. Denton: University of North Texas Press.

Lehman, Frank. 2012a. "Music Theory Through the Lens of Film." *Journal of Film Music* 5(1–2): 179–198.

———. 2012b. "Reading Tonality Through Film: Transformational Hermeneutics and the Music of Hollywood." PhD dissertation, Harvard University.

———. 2013a. "Hollywood Cadences: Music and the Structure of Cinematic Expectation." *Music Theory Online* 19(4). https://mtosmt.org/issues/mto.13.19.4/mto.13.19.4.lehman.html.

———. 2013b. "Transformational Analysis and the Representation of Genius in Film Music." *Music Theory Spectrum* 35(1): 1–22.

———. 2017. "Manufacturing the Epic Score: Hans Zimmer and the Sounds of Significance." In *Music in Epic Film: Listening to the Spectacle*, edited by Stephen C. Meyer, 27–55. New York: Routledge Press.

———. 2018. *Hollywood Harmony: Musical Wonder and the Sound of Cinema*. New York: Oxford University Press.

Leinberger, Charles. 2002. "Thematic Variation and Key Relationships: Charlotte's Theme in Max Steiner's Score for Now, Voyager." *Journal of Film Music Studies* 1: 63–77.

Levarie, Siegmund. 1978. "Key Relations in Verdi's Un Ballo in Maschera." *19th Century Music* 2(2): 143–147.

Levinson, Jerrold. 1996. "Film Music and Narrative Agency." In *Post-Theory: Reconstructing Film Studies*, edited by David Bordwell and Nöel Carroll, 248–282. Madison: University of Wisconsin Press.

Lewin, David. 1986. "Music Theory, Phenomenology, and Modes of Perception." *Music Perception* 3(4): 327–392.

Livio, Mario. 2002. *The Golden Ratio: The Story of Phi, the World's Most Astonishing Number*. New York: Broadway Books.

Marco, Guy A., and Siegmund Levarie. 1979. "On Key Relations in Opera." *19th Century Music* (3)1: 83–89.

Marcozzi, Rudy. 1992. "The Interaction of Large-Scale Harmonic and Dramatic Structure in the Verdi Operas Adapted from Shakespeare." PhD dissertation, Indiana University.

Max Steiner Collection. MSS 1547, The Max Steiner Collection; Film Music Archives. L. Tom Perry Special Collections, Harold B. Lee Library, Brigham Young University.

McCreless, Patrick. 1982. *Wagner's Siegfried: Its Drama, History, and Music*. Ann Arbor: UMI Research Press.

———. 1990. "Schenker and Chromatic Tonicization: A Reappraisal." In *Schenker Studies*, edited by Heidi Siegel, 125–145. Cambridge: Cambridge University Press.

———. 1996. "An Evolutionary Perspective on Nineteenth-Century Semitonal Relations." In *The Second Practice of Nineteenth-Century Tonality*, edited by William Kinderman and Harald Krebs, 87–113. Lincoln: University of Nebraska Press.

———. 2010. "Shostakovich's Politics of D Minor and Its Neighbors, 1931–1949." In *Shostakovich Studies* 2, edited by Pauline Fairclough. Cambridge: Cambridge University Press.

McQuiston, Kate. 2017. "Some Assembly Required: Hybrid Scores in *Moonrise Kingdom* and *The Grand Budapest Hotel*." In *The Routledge Companion to Screen Music and Sound*, edited by Miguel Mera, Ronald Sadoff, and Ben Winters, 477–493. New York: Routledge Press.

Medina-Gray, Elizabeth. 2019. "Analyzing Modular Smoothness in Video Game Music." *Music Theory Online* 25(3). https://www.mtosmt.org/issues/mto.19.25.3/mto.19.25.3.medina.gray.html.

———. 2021. "Interacting with Soundscapes: Music, Sound Effects and Dialogue in Video Games." In *The Cambridge Companion to Video Game Music*, edited by Melanie Fritsch and Tim Summers, 176–192. Cambridge: Cambridge University Press.

Monahan, Seth. 2013. "Action and Agency Revisited." *Journal of Music Theory* 57(2): 321–371.

Monelle, Raymond. 2000. *The Sense of Music*. Princeton, NJ: Princeton University Press.

Motazedian, Táhirih. 2016. "To Key or Not to Key: Tonal Design in Film Music." PhD dissertation, Yale University.

———. Forthcoming. "Tonal Analysis of the Integrated Soundtrack: Music, Sound, and Dialogue in Baby Driver." In *Music Analysis and Film: Studying the Score: A Routledge Handbook on Music and Screen Media*, edited by Frank Lehman. New York: Routledge Press.

Murch, Walter. 1995. "Sound Design: The Dancing Shadow." In *Projections 4: Film-makers on Film-making*, edited by John Boorman, Tom Luddy, David Thompson, and Walter Donohue, 237–251. Boston: Faber and Faber.

Murphy, Scott. 2006. "The Major Tritone Progression in Recent Hollywood Science Fiction Films." *Music Theory Online* 12(2). https://mtosmt.org/issues/mto.06.12.2/mto.06.12.2.murphy.html.

———. 2014a. "A Pop Music Progression in Recent Popular Movies and Movie Trailers." *Music, Sound, and the Moving Image* 8(2): 141–162.

———. 2014b. "Scoring Loss in Some Recent Popular Film and Television." *Music Theory Spectrum* 36(2): 295–314.

———. 2014c. "Transformational Theory and the Analysis of Film Music." In *The Oxford Handbook of Film Music Studies*, edited by David Neumeyer, 471–499. New York: Oxford University Press.

———. 2023 forthcoming. "An Eightfold Taxonomy of Harmonic Progressions, and Its Application to Triads Related by Major Third and Their Significance in Recent Screen Music." *Journal of Music Theory* 67(1).

Nattiez, Jean-Jacques. 1990. *Music and Discourse: Toward a Semiology of Music*, translated by Carolyn Abbate. Princeton, NJ: Princeton University Press.

Nehamas, Alexander. 1986. "What an Author Is." *Journal of Philosophy* 83(11): 685–691.

Neumeyer, David. 1998. "Tonal Design and Narrative in Film Music: Bernard Herrmann's *A Portrait of Hitch* and *The Trouble With Harry*." *Indiana Theory Review* 19: 87–123.

———. 2009. "Diegetic/Nondiegetic: A Theoretical Model." *Music and the Moving Image* 2(1): 26–39.

———. 2015. *Meaning and Interpretation of Music in Cinema*. With contributions by James Buhler. Bloomington: Indiana University Press.

Neumeyer, David, and James Buhler. 2001. "Analytical and Interpretive Approaches to Film Music (I): Analyzing the Music." In *Film Music: Critical Approaches*, edited by Kevin Donnelly, 16–38. Edinburgh: Edinburgh University Press.

Nobile, Drew. 2020. "Double-Tonic Complexes in Rock Music." *Music Theory Spectrum* 42(2): 207–226.

O'Meara, Jennifer. 2014. "A Shared Approach to Familial Dysfunction and Sound Design: Wes Anderson's Influence on the Films of Noah Baumbach." In *The Films of Wes Anderson: Critical Essays on an Indiewood Icon*, edited by Peter C. Kunze, 109–124.

Orgeron, Devin. 2007. "La Camera-Crayola: Authorship Comes of Age in the Cinema of Wes Anderson." *Cinema Journal* 46(2): 40–65.

Platte, Nathan. 2011. "Music for Spellbound (1945): A Contested Collaboration." *Journal of Musicology* 28(4): 418–463.

Poluyko, Kristen. 2011. "Alternative Music: Jazz and the Performative Resignification of Identity in Anthony Minghella's *The Talented Mr. Ripley*." *Journal of Men's Studies* 19(1): 19–36.

Powrie, Phil, and Robynn Stilwell, eds. 2006. *Changing Tunes: The Use of Pre-Existing Music in Film*. Aldershot: Ashgate Publishing.

Reyland, Nicholas. 2015. "Corporate Classicism and the Metaphysical Style." *Music, Sound, and the Moving Image* 9(2): 115–130.

Richards, Mark. 2017. "Tonal Ambiguity in Popular Music's Axis Progressions." *Music Theory Online* 23(3). https://mtosmt.org/issues/mto.17.23.3/mto.17.23.3.richards.html.

Rodman, Ronald. 1998. "'There's No Place Like Home:' Tonal Closure and Design in the Wizard of Oz." *Indiana Theory Review* 19: 125–43.

———. 2000. "Tonal Design and the Aesthetic of Pastiche in Herbert Stothart's Maytime." In *Music and Cinema*, edited by James Buhler, Caryl Flinn, and David Neumeyer, 187–206. Middletown, CT: Wesleyan University Press.

———. 2010. *Tuning in: American Narrative Television Music*. New York: Oxford University Press.

———. 2011. "The Operatic Stothart: Leitmotifs and Tonal Organization in Two Versions of 'Rose Marie.'" *Journal of Film Music* 4(1): 5–19.

Rosar, William. 2006. "Music for Martians: Schillinger's Two Tonics and Harmony of Fourths in Leith Steven's Score for War of the Worlds." *Journal of Film Music* 1(4): 395–438.

Rudolph, Pascal. 2022. "The Musical Idea Work Group: Production and Reception of Pre-existing Music in Film." *Twentieth-Century Music* 20(2): 1–21.

Sagan, Carl. 1973. "Sagan." In *Life Beyond Earth and the Mind of Man*, edited by Richard Berendzen, 5–14. Washington, DC: NASA Scientific and Technical Information Office.

Salzman, Eric. 1974. *Twentieth-Century Music: An Introduction*, 2nd edition. Englewood Cliffs, NJ: Prentice-Hall, Inc.

Schatz, Thomas. 2010. *The Genius of the System: Hollywood Filmmaking in the Studio Era*. Minneapolis: University of Minnesota Press.

Schneller, Tom, and Táhirih Motazedian. 2017. "Tugging at Heartstrings: Bittersweet Harmonies in the Classic Hollywood Love Theme." *Journal of Film Music* 10(2): 7–48.

Schultz, Rob. 2012. "Tonal Pairing and the Relative-Key Paradox in the Music of Elliott Smith." *Music Theory Online* 18(4). https://mtosmt.org/issues/mto.12.18.4/mto.12.18.4.schultz.html.

Sergi, Gianluca. 2016. "Organizing Sound: Labour Organizations and Power Struggles that Helped Define Music and Sound in Hollywood." In *The Palgrave Handbook of Sound Design and Music in Screen Media*, edited by Liz Greene and Danijela Kulezic-Wilson, 43–56. London: Palgrave Macmillan.

Shapiro, Ari (host). 2017, February 20. "Song Exploder: 'Moonlight' Composer Describes Process" [Audio podcast transcript]. In *All Things Considered*. NPR. www.npr.org/2017/02/20/516292253/song-exploder-moonlight-composer-describes-process.

Smith, Jeff. 2009. "Bridging the Gap: Reconsidering the Border between Diegetic and Nondiegetic Music." *Music and the Moving Image* 2(1): 1–25.

Spicer, Mark. 2017. "Fragile, Emergent, and Absent Tonics in Pop and Rock Songs." *Music Theory Online* 23(2). https://mtosmt.org/issues/mto.17.23.2/mto.17.23.2.spicer.html.

Stein, Deborah. 1985. *Hugo Wolf's Lieder and Extensions of Tonality*. Ann Arbor: UMI Research Press.

Stilwell, Robynn. 2007. "The Fantastical Gap between Diegetic and Nondiegetic." In *Beyond the Soundtrack: Representing Music in Cinema*, edited by Daniel Goldmark, Lawrence Kramer, and Richard Leppert, 187–202. Berkeley: University of California Press.

Taruskin, Richard. 1996. *Stravinsky and the Russian Traditions*. Berkeley: University of California Press.

———. 2000. *Defining Russia Musically: Historical and Hermeneutical Essays*. Princeton, NJ: Princeton University Press.

———. 2004. "The Poietic Fallacy." *The Musical Times* 145(1886): 7–34.

———. 2009. *On Russian Music*. Berkeley: University of California Press.

Taylor, Richard, Adam Micolich, and David Jonas. 1999. "Fractal Analysis of Pollock's Drip Paintings." *Nature* 399(6735): 422.

Temperley, David. 1999. "The Question of Purpose in Music Theory: Description, Suggestion, and Explanation." *Current Musicology* 66: 66–85.

Thompson, Kristin, and David Bordwell. 2014, March 26. "THE GRAND BUDAPEST HOTEL: Wes Anderson takes the 4:3 challenge." www.davidbordwell.net/blog/2014/03/26/the-grand-budapest-hotel-wes-anderson-takes-the-43-challenge/.

Van Buskirk, Eliot. 2015, May 6. "The Most Popular Keys of All Music on Spotify." *Spotify Insights*. https://web.archive.org/web/20160201150029/https://insights.spotify.com/us/2015/05/06/most-popular-keys-on-spotify/.

Verba, Emily. 2012. "The Golden Ratio in Time-based Media." *Journal of Arts and Humanities* 1(1): 56–68.

Watts, Catrin. 2018. "Blurred Lines: The Use of Diegetic and Nondiegetic Sound in Atonement (2007)." *Music and the Moving Image* 11(2): 23–36.

Weber, Gottfried. 1817 (1851). *The theory of musical composition, treated with a view to a naturally consecutive arrangement of topics*. London: R. Cocks and Co.

White, Terri. 2017, June 19. "Baby Driver Review." *Empire*. www.empireonline.com/movies/reviews/baby-driver-review/.

"Willie Nelson Albums Discography." N.d. Wikipedia. https://en.wikipedia.org/wiki/Willie_Nelson_albums_discography#1970s, accessed 1 October 2021.

Wimsatt, W. K., Jr., and M. C. Beardsley. 1946. "The Intentional Fallacy." *The Sewanee Review* 54(3): 468–488.

Winters, Ben. 2010. "The Non-Diegetic Fallacy: Film, Music, and Narrative Space." *Music & Letters* 91(2): 224–244.

———. 2012. "It's All Really Happening: Sonic Shaping in the Films of Wes Anderson." In *Music, Sound and Filmmakers: Sonic Style in Cinema*, edited by James Wierzbicki, 45–60. New York: Routledge Press.

INDEX

Figures and tables are indicated with fig. or tab. following the page number. Footnotes are indicated with n between the page number and the note number. Preexisting musical works are indexed under the name of the composer or performer(s).

Abbate, Carolyn, 4–5
acousmatic sound, 29, 109
Adorno, Theodor, 154–155
Amadeus (1984): associative tonality in, 42–45; cadential frustration in, 43, 44*fig.*, 45, 47–52, 48*fig.*, 49*tab.*, 50*fig.*, 51*fig.*, 120; functional tonality in, 43–44, 45–47, 46*fig.*, 47*fig.*, 52*fig.*; parallel key relationships in, 42; synopsis, 41–42, 52*fig.*; tonal agency in, 42–43
American Graffiti (1973), 153
analysis of film music: approaches to, 2–7, 11–19, 55, 142–144, 154–156, 157–161; esthesic vs. poietic, 9–12, 150; selectivity in, 11, 13, 18, 55; subjectivity of, 10, 155
Anderson, Wes, 35, 41, 98–99, 100n17, 104–105, 109n1, 146, 147. See also *Darjeeling Limited, The* (2007); *Fantastic Mr. Fox* (2009); *Grand Budapest Hotel, The* (2014); *Moonrise Kingdom* (2012)
Annie Hall (1977), 143
associative tonality: defined, 19–20, 24; in *Amadeus*, 42–45; approaches to analyzing, 13–14, 158, 160–161; in *Baby Driver*, 128, 129*fig.*, 130, 132, 134, 138–139, 141, 141*fig.*; in *The Darjeeling Limited*, 39–41; in *Emma*, 75, 75*tab.*; in *The English Patient*, 29–33, 109–110; in *Fantastic Mr. Fox*, 77–81, 101, 111–113, 112*tab.*; in *The Graduate*, 57–58, 113–114; in *The Grand Budapest Hotel*, 99–100, 100*tab.*, 101–103, 110–111, 112*tab.*; in *Moonlight*, 65–67; in *Moonrise Kingdom*, 35–36, 113; in *Persuasion*, 59–60, 60*fig.*, 61*fig.*, 62; in *The Royal Tenenbaums*, 69–70, 70*tab.*, 71*tab.*, 72; and sound effects, 109–114, 112*tab.*; in *The Talented Mr. Ripley*, 84–85, 85*tab.*, 86*tab.*, 87–93, 87*fig.*, 90*fig.*, 91*fig.*, 95–97, 106–107
At Home with Amy Sedaris, 53–54
Atonement (2007), 121–122, 154
audibility, 4–6, 155
auteurism, 7, 104, 145–147
authorial intent. *See* intentionality
authorship, 7–8, 9–10n21, 150

Baby Driver (2017): associative tonality in, 128, 129*fig.*, 130, 132, 134, 138–139, 141, 141*fig.*; functional tonality in, 138, 141–142; intertextual tonality in, 128, 129*fig.*, 130, 134, 141; parallel key relationships in, 134, 140–141, 141*fig.*; preexisting music in, 128, 129*fig.*, 130–132, 133–134, 135–138*fig.*, 138–142, 140*tab.*, 146–147, 149; sound effects in, 128, 129*fig.*, 130–134, 133*tab.*, 135–138*fig.*, 142; synopsis, 128; tonal

Baby Driver (2017) *(continued)*
  agency in, 130–131, 133–134; tragic-to-triumphant arcs in, 134, 140–141, 141*fig.*; transposition in, 139–140, 140*tab.*, 149; Wright on, 141, 146–147
Bach, J. S.: B-A-C-H motif, 20, 155; Fantasia in C Minor (BWV 562), 120; French Suites, 62–63, 63*tab.*, 160; Goldberg Variations, 16–17, 30, 32–33; *Italian Concerto*, 85, 87–88, 94, 96–97; Sinfonia No. 2 in C minor (BWV 788), 37; *St. Matthew Passion*, 92, 94
Bailey, Robert, 19–20
Battleship Potemkin, The (1925), 104
Beach, David, 3. *See also* tonal design; tonal structure
Beethoven, Ludwig van: Piano Quartet, Op. 16, 92, 93–94; Symphony No. 5, 2, 5, 24; Symphony No. 7, 41, 115, 116*fig.*
Benny and Joon (1993), 121, 122*fig.*
Bob & Earl, "Harlem Shuffle," 131, 139–140, 140*tab.*
Breaking and Entering (2006), 36–37, 114
Bribitzer-Stull, Matthew, 4n4, 5n7, 20, 21n39, 22n41, 42
Bridesmaids (2011), 144
Britell, Nicholas, 65–67, 65*fig.*, 148–149. *See also Moonlight* (2016)
Britten, Benjamin, 33–36, 34*tab.*
Buhler, James, 3n3, 6n11, 7n12, 142

cadential frustration: defined, 25; in *Amadeus*, 43, 44*fig.*, 45, 47–52, 48*fig.*, 49*tab.*, 50*fig.*, 51*fig.*, 120; in *Super Mario Bros* (video game), 151; in *The Talented Mr. Ripley*, 88, 93–94, 95; in *When Harry Met Sally. . .*, 153
Chopin, Frédéric, 59, 61*fig.*, 62–63, 63*tab.*
"chopped and screwed" technique, 66, 148
classical music: analytical approaches to, 2–3, 5, 55, 107; in film (see *Amadeus* [1984]; *Moonrise Kingdom* [2012]; *Persuasion* [1995]; *Talented Mr. Ripley, The* [1999]); harmonic logics of, 3–4, 6n9, 19–20, 23, 24, 154–155; influence on film music, 54–55, 155
closing credits, 22, 33, 37, 62, 70, 100, 115
clubs, music in, 150, 152
Coen brothers, 150
collaboration, 7–8, 147
Commodores, "Easy," 138, 139, 140*tab.*, 141
compilation scoring, 18, 57–58
Cook, Nicholas, 5
Crosby, Bing, "May I?" 90–91, 91*fig.*, 92*fig.*, 94

Darjeeling Limited, The (2007): analysis, 14–15, 39–41, 39*tab.*, 40*fig.*, 114–115, 116*fig.*; synopsis, 38–39
Davis, Miles, "Nature Boy," 94
Delerue, Georges, 78–79, 81
Desplat, Alexandre, 34–36, 77, 78–79, 99. *See also Fantastic Mr. Fox* (2009); *Grand Budapest Hotel, The* (2014); *Moonrise Kingdom* (2012)
Detroit Emeralds, "Baby Let Me Take You," 134, 135–138*fig.*
directional tonality: defined, 24, 81; in *Emma*, 75, 75*tab.*; in *Fantastic Mr. Fox*, 81; in *The Graduate*, 57–58, 155; in *Hidden Figures*, 72–73; in *Moonlight*, 67; in *Persuasion*, 2, 59–60, 60*fig.*, 61*fig.*, 62–64; in *The Royal Tenenbaums*, 69–70, 70*tab.*, 72, 73*tab.*
double-tonic complexes, 3, 18, 21
drastic perception (Abbate), 4–6

Eisler, Hanns, 154–155
Emma (1996), 73–75, 74*tab.*, 75*tab.*
English Patient, The (1996): analysis, 16–18, 16*fig.*, 17*fig.*, 28–33, 109–110, 126, 127*fig.*; Minghella on, 146; synopsis, 27–28
esthesic analysis, 9–12, 150

Fantastic Mr. Fox (2009): analysis, 2, 76–81, 77*fig.*, 80*tab.*, 101, 111–113, 112*tab.*, 121; synopsis, 1, 76
Favourite, The (2018), 118–120
Ferrari, Luc, 119–120
film music, analytical approaches to, 2–7, 11–19, 55, 142–144, 154–156, 157–161
film tonality, 3–4, 11–12, 26, 152–153
Fistful of Dollars, A (1964), 147, 154
flashbacks: in *The Darjeeling Limited*, 115, 116*fig.*; in *The English Patient*, 109–110; in *The Grand Budapest Hotel*, 82, 97–98, 99–100, 100*fig.*, 105–106; in *The Talented Mr. Ripley*, 82, 84–85, 105–106
For a Few Dollars More (1965), 76–77, 147, 154
Friends, 54
functional tonality: defined, 25; in *Amadeus*, 43–44, 45–47, 46*fig.*, 47*fig.*, 52*fig.*; in *Baby Driver*, 138, 141–142; in *Emma*, 75, 75*tab.*; in *Fantastic Mr. Fox*, 78–79, 112, 112*tab.*; in *The Favourite*, 118–120; in *The Graduate*, 58, 117; in *The Grand Budapest Hotel*, 124; listening for, 159; in *Lost in London*, 117–118. *See also* cadential frustration

INDEX

Gardiner, Boris, 66, 67
Gilbert & Sullivan, "We're Called Gondolieri," 91, 92*fig.*
gnostic perception (Abbate), 4–6, 11–12
Golden Ratio, 104, 105*fig.*, 107, 150
Google Chat, 151, 152*tab.*
Gorbman, Claudia, 108, 145, 156
Graduate, The (1967), 13, 56–58, 113–114, 117, 155
Grand Budapest Hotel, The (2014): associative tonality in, 99–100, 100*tab.*, 101–103, 110–111, 112*tab.*; compared to *The Talented Mr. Ripley*, 105–107; diegetic music in, 102, 112*tab.*; and the Golden Ratio, 104, 105*fig.*, 107, 150; intertextual tonality in, 101; keys as narrative advancement device in, 100–101, 101*tab.*; meta keys in, 102–104, 103*fig.*, 105*fig.*, 150; preexisting music in, 101*tab.*, 102, 153; sound effects in, 99, 100, 108–109, 110–111, 112*tab.*, 123–124, 124*fig.*, 125*fig.*; synopsis, 97–98; tonal symmetry in, 82, 99–100, 103–104, 103*fig.*, 105*fig.*, 106, 106*fig.*, 107, 156; transposition in, 102
Grusin, Dave, 57. See also *Graduate, The* (1967)

harmonic mixing, 150
Harrelson, Woody, 117–118
"Heartstring Schema" (chord progression), 67
Herrmann, Bernard, 143. See also *Psycho* (1960)
hexatonic pole relationships, 54, 79
Hidden Figures (2016), 72–73

I Love You Phillip Morris (2009), 144
intentionality, 9–12, 145–150, 151
intertextual tonality: defined, 21; in *Baby Driver*, 128, 129*fig.*, 130, 134, 141; in *Breaking and Entering*, 37; in *The English Patient*, 32–33; in *Fantastic Mr. Fox*, 77; in *The Grand Budapest Hotel*, 101; in *Moonrise Kingdom*, 34–36; in *The Talented Mr. Ripley*, 84–85, 87–88, 89–93, 92*fig.*, 96–97, 156

Jenkins, Barry, 65, 148. See also *Moonlight* (2016)
Jon Spencer Blues Explosion, "Bellbottoms," 128, 129*fig.*, 130, 141, 146
jump-cut tonality, 26, 52–54

key constellations, 10–11, 21, 156
Kinks, The, "Powerman," 41
Kmet, Nicholas, 8

Laing, Heather, 83–84
Lanthimos, Yorgos, 118. See also *Favourite, The* (2018)
Lehman, Frank, 11n26, 19n35, 24n46, 25, 25n50, 150
leitmotifs, 20, 65, 65*fig.*
Leittonwechsel transformations, 17–18, 23, 29–30, 110
listener perception, 4–6, 12, 26, 55, 152, 155–156
long-range tonality, 4–6, 153, 154–155
Lost in London (2017), 117–118

Marianelli, Dario, 121, 123. See also *Atonement* (2007)
mélomanes, 145–147
Mendelssohn, Felix, "Wedding March," 113–114, 117
metadiegetic sound, 12, 13, 47, 122–123, 134
meta keys: defined, 25–26; in *The Grand Budapest Hotel*, 102–104, 103*fig.*, 105*fig.*; in *The Royal Tenenbaums*, 70
Minghella, Anthony, 17, 37, 83, 145–146, 148. See also *Breaking and Entering* (2006); *English Patient, The* (1996); *Talented Mr. Ripley, The* (1999)
mise-en-bande, 6–7, 11, 55, 142. See also soundtracks
mise-en-scène, 6, 98, 104–105
modulations, 24, 37, 80, 80*tab.*, 102, 154, 159. See also jump-cut tonality
monotonality, 3–4, 6, 21, 154
montage sequences: in *Amadeus*, 43; in *Baby Driver*, 138–139, 140*tab.*; in *Breaking and Entering*, 37, 114; in *Fantastic Mr. Fox*, 111–112, 112*tab.*; in *The Graduate*, 57; in *The Grand Budapest Hotel*, 102, 111; in *The Royal Tenenbaums*, 69, 70*tab.*; in *The Talented Mr. Ripley*, 87–90
Moonlight (2016): analysis, 65–68, 65*fig.*, 124–125, 148–149; Britell on, 148–149; compared to *Persuasion*, 67–68; synopsis, 64–65
Moonrise Kingdom (2012): analysis, 34–36, 36*fig.*, 113; Britten's works in, 34–35, 34*tab.*; synopsis, 33–34
Morricone, Ennio, 76–77, 147–148, 154
Mothersbaugh, Mark, 35, 69–70, 70*tab.*, 71*tab.*, 73*tab.*. See also *Royal Tenenbaums, The* (2001)
Mozart, Wolfgang Amadeus: *Abduction from the Seraglio*, 43, 48, 49; *Don Giovanni*, 45; Klavierstück K. 33B in F major, 42; *The Magic Flute*, 45, 51; *The Marriage of Figaro*, 43–44,

Mozart, Wolfgang Amadeus *(continued)* 49–50, 51*fig.*; Mass in C minor, K. 427, 49, 50*fig.*; Piano Concerto No. 20, 45, 51–52; *Requiem*, 44–45, 51, 52*fig.*; Serenade for Winds, K. 361, 47, 48*fig*. See also *Amadeus* (1984)
Murphy, Scott, 6, 19n35, 23, 25n50

NBC chime, 20, 151
Nelson, Willie, 117–118
Neumeyer, David, 3n3, 7, 7n12, 83, 122n10, 142
Nichols, Mike, 57. See also *Graduate, The* (1967)

Office, The, 53
On Golden Pond (1981), 153
opening credits, 69, 98–99, 100, 100*tab.*, 122
opera, 19–20, 23–24. See also *Amadeus* (1984); *Moonrise Kingdom* (2012); *Talented Mr. Ripley, The* (1999); Wagner, Richard

parallel key relationships: defined, 22, 24; in *Amadeus*, 42; in *Baby Driver*, 134, 140–141, 141*fig.*; in *Fantastic Mr. Fox*, 78, 80, 81; in *Moonrise Kingdom*, 35–36; in *The Royal Tenenbaums*, 69–70, 70*tab.*, 71*tab.*, 72; in *The Talented Mr. Ripley*, 88–91, 92*fig.*
Parker, Charlie, "Ko-Ko," 92, 94
Pergolesi, Giovanni Battista, *Stabat Mater*, 42, 46–47, 46*fig.*, 47*fig.*, 51
Persuasion (1995): analysis, 2, 59–60, 60*fig.*, 61*fig.*, 62–64, 63*fig.*, 160; compared to *Moonlight*, 67–68; synopsis, 1, 58–59
pitched dialogue, 144
poietic analysis, 9–10, 150
Pollock, Jackson, 149–150
Portman, Rachel, 74–75, 121, 122*fig*. See also *Benny and Joon* (1993); *Emma* (1996)
Poster, Randall, 146
preexisting music: in *Atonement*, 123; in *Baby Driver*, 128, 129*fig.*, 130–132, 133–134, 135–138*fig.*, 138–142, 140*tab.*, 149; in *Breaking and Entering*, 37; in *The Darjeeling Limited*, 39*tab.*, 41, 115, 116*fig.*; in *The English Patient*, 16–17, 29–33, 110; in *Fantastic Mr. Fox*, 78–81; in *The Favourite*, 119–120; in *The Graduate*, 57–58, 113–114, 117; in *The Grand Budapest Hotel*, 101*tab.*, 102, 153; in *Lost in London*, 117–118; in *Moonlight*, 66–67; in *Moonrise Kingdom*, 34–36, 34*tab.*; in *Persuasion*, 2, 59, 61*fig.*, 62–63, 63*tab.*; in *The Royal Tenenbaums*, 70*tab.*, 71*tab.*; strategies for identifying, 160; in *The Talented Mr. Ripley*, 83, 85–86, 87–89, 87*fig.*, 87*tab.*, 90–95, 91*fig.*, 96–97; transposition of, 2, 57–58, 62–63, 63*tab.*, 66–67, 80–81; in *Wall Street*, 126–127. See also *Amadeus* (1984)
Pretty Woman (1990), 143
Price, Steven, 134. See also *Baby Driver* (2017)
Psycho (1960), 143

radio, 150, 152
reader-based analysis. *See* esthesic analysis
relative key relationships: defined, 22–23; in *Moonrise Kingdom*, 35–36; in *The Talented Mr. Ripley*, 85, 85*fig.*, 87–88, 96–97, 96*fig.*, 97*fig.*
Rodgers and Hart, "My Funny Valentine," 88–89
Rodman, Ronald, 6, 9, 11n25, 21
Rollins, Sonny, "Tenor Madness," 92
Royal Tenenbaums, The (2001): analysis, 69–70, 70*tab.*, 71*tab.*, 73*tab.*, 108–109; Anderson's approach to music in, 146; synopsis, 68–69

Sabaneev, Leonid, 152
Sagan, Carl, 156
Salieri, Antonio. See *Amadeus* (1984)
Sams, Jeremy, 59, 61*fig*. See also *Persuasion* (1995)
Schumann, Robert, Piano Quintet, 120
Sedaris, Amy, 53–54
Shostakovich, Dmitri, 5–6, 20, 21
Simon & Garfunkel: "Baby Driver," 134, 139, 141, 149; "Mrs. Robinson," 113–114, 117; "Scarborough Fair," 13; "The Sound of Silence," 57–58, 117
singular keys: defined, 22; in *The Darjeeling Limited*, 40–41; in *Fantastic Mr. Fox*, 80; in *The Royal Tenenbaums*, 70, 72; in *The Talented Mr. Ripley*, 95
sound advance, 88, 89
sound effects:
—CATEGORIES AND ANALYTICAL APPROACHES: defined, 108; associative, 109–114, 112*tab.*; concordant, 125–128, 127*fig.*; distance from musical cues, 153; listening for, 142–144, 160–161; musicalized, 120–123, 122*fig.*; opening, 108–109; stepwise, 114–115, 116*fig.*; tonally functional, 117–120; transposed, 123–125, 124*fig.*, 125*fig.*, 126
—EXAMPLES: *Annie Hall*, 143; app design, 151, 152*tab.*; *Atonement*, 121–123; *Baby Driver*, 128, 129*fig.*, 130–134, 133*tab.*, 135–138*fig.*, 142; *Breaking and Entering*, 114; *The Darjeeling Limited*, 114–115, 116*fig.*; *The English Patient*,

28–29, 109–110, 126, 127*fig.*; *Fantastic Mr. Fox*, 111–113, 112*tab.*, 121; *The Favourite*, 119–120; *On Golden Pond*, 153; *The Graduate*, 113–114; *The Grand Budapest Hotel*, 99, 100, 108–109, 110–111, 112*tab.*, 123–124, 124*fig.*, 125*fig.*; *Lost in London*, 117–118; *Moonlight*, 66–67, 124–125; *Moonrise Kingdom*, 113; *Psycho*, 143; *Pretty Woman*, 143; *The Royal Tenenbaums*, 69, 108–109; *30 Rock*, 143; *2001: A Space Odyssey*, 144; video games, 151; *Wall Street*, 126–127; *When Harry Met Sally. . .*, 153

sound lag, 46, 89

soundtracks: albums of, 160; analytical approaches to, 2–7, 11–19, 55, 142–144, 154–156, 157–161; components of, 3, 12–13; composite authorship of, 7–8; as compositions, 6–8; gaps in, 5–6, 152–153

Spaghetti Westerns, 76–78, 147, 154

Steiner, Max, 148

stepwise relationships: defined, 23–24; in *Breaking and Entering*, 37, 114; in *The Darjeeling Limited*, 39–40, 40*fig.*, 115, 116*fig.*

structural tonality. *See* tonal structure

subjectivity of analysis, 10, 155

Super Mario Bros (1985 video game), 151

Talented Mr. Ripley, The (1999): associative tonality in, 84–85, 85*tab.*, 86*tab.*, 87–93, 87*fig.*, 90*fig.*, 91*fig.*, 95–97, 106–107; cadential frustration in, 88, 93–94, 95; compared to *The Grand Budapest Hotel*, 105–107; diegetic music in, 84–85, 86*fig.*, 87–89, 87*fig.*, 90–95, 91*fig.*; intertextual tonality in, 84–85, 87–88, 89–93, 92*fig.*, 96–97, 156; parallel key relationships in, 88–91, 92*fig.*; preexisting music in, 83, 85–86, 87–89, 87*fig.*, 87*tab.*, 90–95, 91*fig.*, 96–97; relative key relationships in, 85, 85*fig.*, 87–88, 96–97, 96*fig.*, 97*fig.*; singular keys in, 95; synopsis, 82–83; tonal agency in, 88, 89–91; tonal codas in, 96, 97*fig.*; tonal pairings in, 85, 85*fig.*, 87–88, 96–97, 96*fig.*, 97*fig.*; tonal symmetry in, 82, 84, 96, 97*fig.*, 106–107, 106*fig.*; transposition in, 93

"Tarnhelm" progression, 54

Tchaikovsky, Pyotr Ilyich, *Eugene Onegin*, 84–85, 86*fig.*, 96, 107

television, 52–54

30 Rock, 52–53, 143

through-composed scores, 18–19

tonal agency: defined, 20–21; in *Amadeus*, 42–43; in *Baby Driver*, 130–131, 133–134; and cadential frustration, 25; in *The Darjeeling Limited*, 40–41; in *Fantastic Mr. Fox*, 78–79; in *The Talented Mr. Ripley*, 88, 89–91

tonal codas: defined, 22; in *Breaking and Entering*, 37; in *The English Patient*, 33; in *The Talented Mr. Ripley*, 96, 97*fig.*

tonal continuity, 150–153

tonal design, 3–4, 10, 11, 19–26, 150–154

tonal graphs: defined, 14–15, 18; *The Darjeeling Limited*, 14*fig.*; *The English Patient*, 16*fig.*; *The Grand Budapest Hotel*, 103*fig.*, 105*fig.*; process of creating, 158; *The Talented Mr. Ripley*, 96*fig.*

tonal intertextuality. *See* intertextual tonality

tonality by assertion, 126, 159

tonal pairings: defined, 21–22; in *The English Patient*, 29–31, 32–33; in *Moonrise Kingdom*, 35–36; in *The Talented Mr. Ripley*, 85, 85*fig.*, 87–88, 96–97, 96*fig.*, 97*fig.*

tonal scores, 12–14, 157–161

tonal staffs: defined, 15–16; *The Darjeeling Limited*, 40*fig.*; *The English Patient*, 16*fig.*, 17*fig.*; process of creating, 158; *The Talented Mr. Ripley*, 97*fig.*

tonal structure, 3–4, 11, 154–155

tonal symmetry: defined, 26; in classical music, 107; in *The Grand Budapest Hotel*, 82, 99–100, 103–104, 103*fig.*, 105*fig.*, 106, 106*fig.*, 107, 156; in *The Talented Mr. Ripley*, 82, 84, 96, 97*fig.*, 106–107, 106*fig.*

tonal winks: defined, 20; in *Fantastic Mr. Fox*, 79; in Google sound effects, 151, 152*tab.*; in *Moonrise Kingdom*, 35, 36*fig.*, 113

tragic-to-triumphant arcs: defined, 24; in *Baby Driver*, 134, 140–141, 141*fig.*; in *Fantastic Mr. Fox*, 81; in *Moonrise Kingdom*, 35; in *The Royal Tenenbaums*, 69–70, 70*tab.*, 72

transposition: and associative tonality, 20; in *Baby Driver*, 139–140, 140*tab.*, 149; and directional tonality, 81; in *Emma*, 74–75, 74*tab.*, 75*tab.*; in *The English Patient*, 29; in *Fantastic Mr. Fox*, 77–78, 80–81, 80*tab.*; in *The Graduate*, 57–58; in *The Grand Budapest Hotel*, 102; in *Hidden Figures*, 72–73; and intentionality, 148–149; listening for, 159–160; in *Moonlight*, 65–67, 148–149; in *Persuasion*, 62–63, 63*fig.*; of preexisting music, 2, 20, 57–58, 62–63, 63*fig.*, 66–67, 80–81, 139–140, 140*tab.*, 148–149, 159–160; in *The Royal Tenenbaums*, 70; of sound effects, 123–125, 126, 127*fig.*; in *The Talented Mr. Ripley*, 93

T. Rex, "Debora," 140–141, 140*tab.*, 149
2001: A Space Odyssey, 104, 144

video games, 151
Vivaldi, Antonio: Concerto for Lute and Plucked Strings, 101*tab.*, 102; *Stabat Mater*, 88, 95

Wagner, Richard, 5, 20, 23–24, 55, 79, 115n7, 155
Wall Street (1987), 126–127

When Harry Met Sally... (1989), 153
Wright, Edgar, 141, 146–147. See also *Baby Driver* (2017)

Yared, Gabriel, 17, 29, 33, 37, 84, 146, 148. See also *Breaking and Entering* (2006); *English Patient, The* (1996); *Talented Mr. Ripley, The* (1999)

Zimmer, Hans, 8, 73n13

www.ingramcontent.com/pod-product-compliance
Lightning Source LLC
Chambersburg PA
CBHW030654230426
43665CB00011B/1092